Solar Power Generation

The Power Generation Series

Paul Breeze – Coal-Fired Generation, ISBN 13: 9780128040065

Paul Breeze – Gas-Turbine Fired Generation, ISBN 13: 9780128040058

Paul Breeze – Solar Power Generation, ISBN 13: 9780128040041

Paul Breeze – Wind Power Generation, ISBN 13: 9780128040386

Paul Breeze – Fuel Cells, ISBN 13: 9780081010396

Paul Breeze – Energy from Waste, ISBN 13: 9780081010426

Paul Breeze – Nuclear Power, ISBN 13: 9780081010433

Paul Breeze – Electricity Generation and the Environment, ISBN 13: 9780081010440

Solar Power Generation

Paul Breeze

AMSTERDAM • BOSTON • HEIDELBERG • LONDON
NEW YORK • OXFORD • PARIS • SAN DIEGO
SAN FRANCISCO • SINGAPORE • SYDNEY • TOKYO

Academic Press is an imprint of Elsevier

Academic Press is an imprint of Elsevier
125 London Wall, London EC2Y 5AS, UK
525 B Street, Suite 1800, San Diego, CA 92101-4495, USA
50 Hampshire Street, 5th Floor, Cambridge, MA 02139, USA
The Boulevard, Langford Lane, Kidlington, Oxford OX5 1GB, UK

British Library Cataloguing-in-Publication Data
A catalogue record for this book is available from the British Library

Library of Congress Cataloging-in-Publication Data
A catalog record for this book is available from the Library of Congress

ISBN: 978-0-12-804004-1

For Information on all Academic Press publications
visit our website at http://www.elsevier.com/

Working together
to grow libraries in
developing countries

www.elsevier.com • www.bookaid.org

Publisher: Joe Hayton
Acquisition Editor: Lisa Reading
Editorial Project Manager: Peter Jardim
Production Project Manager: Anusha Sambamoorthy
Cover Designer: MPS

Typeset by MPS Limited, Chennai, India

CONTENTS

CHAPTER 1

An Introduction to Solar Power

Solar power is the most important energy resource for life on Earth. The energy in sunlight has driven the evolution of life upon our planet from the earliest tiny organisms through to the plants that have provided food for higher organisms, and eventually for the human race. As a consequence, solar energy is responsible for many of our common energy sources. All the biomass upon the Earth has been created using energy from the Sun to drive photosynthesis, capturing carbon dioxide from the atmosphere and using it to produce organic compounds. The resulting material includes both biomass growing on the Earth today and the fossil fuels that are the remains of ancient biomass buried within the Earth over time, fuels that are now being burned to generate electricity and release the captured carbon dioxide back into the atmosphere.

From the perspective of energy sources, the importance of solar energy is wider still. Sunlight is the heat source that drives the Earth's weather systems. It evaporates the water that generates rainfall, and is therefore responsible for hydropower. The Sun also provides most of the energy that drives the global winds, so it is responsible for wind power. Wave power, a product of the Earth's winds, is indirectly a product of solar power too, as is ocean thermal power. In fact, with the exceptions of nuclear, tidal, and geothermal power, all the major sources of electricity on Earth can be directly or indirectly linked back to the Sun.

While solar energy is responsible for all these exploitable forms of energy, sunlight can also be utilized directly to generate electricity. This can be carried out in two ways. The conceptually simplest method is to use the heat energy contained in solar radiation as a heat source, collecting the Sun's rays and capturing the heat so that it can be used to drive a heat engine such as a steam or gas turbine. This type of power generation has a long history, and today solar thermal power stations are being built in many parts of the world. The second important way

Solar Power Generation. DOI: http://dx.doi.org/10.1016/B978-0-12-804004-1.00001-4

of exploiting solar energy to produce electricity is in a solar or photovoltaic cell. The latter is a solid-state device, closely related to the transistor or microchip, that can absorb sunlight and turn the absorbed light energy into electrical energy. Solar cells use a different part of the solar spectrum from solar thermal power stations, relying on higher energy and shorter wavelength radiation, whereas solar thermal plants use longer wavelength infrared and near-infrared light.

The commercial use of solar energy to generate electricity took off slowly during the last three decades of the 20th century. The main stimuli for the development of both solar thermal and solar photovoltaic technologies were the oil crises of the 1970s. Solar cells soon found a use in applications such as powering satellites and providing remote power. Then a number of nations began funding solar rooftop installation programs, beginning in the early 1990s, to promote the use of solar cells although the cost remained high until the first decade of the 21st century. Since then costs have fallen dramatically and solar cells are becoming cost-effective in a much wider range of applications. For the future, solar cells potentially offer the most cost-effective and simple power-generating technology for providing renewable electric power.

Solar thermal technology has developed much more slowly than solar cells. After an initial burst of activity during the late 1980s and early 1990s, commercial application of the technology was virtually abandoned until the first decade of the 21st century. A number of commercial power plants have since been built but costs still remain relatively high. In spite of that, solar thermal technology has the potential to make an important contribution to global energy production.

The Sun provides an intermittent source of energy at any fixed point on the Earth's surface because it is only available during daylight hours. In addition, the intensity of solar radiation varies with both time of day and atmospheric conditions. As a consequence, solar energy cannot on its own provide a continuous source of electrical power. For small or off-grid applications, solar cells can be backed up with a battery energy storage system that is charged during the day to provide power during hours of darkness. For larger and grid-based solar generation, there is normally an alternative generator available to provide power when solar power is not available.

The amount of electricity generated by solar sources is still relatively small. According to the International Energy Agency (IEA), solar photovoltaic systems generated around 0.85% of global generation in 2013. This was expected to rise to around 1% in 2014. Solar thermal power plants generate about 3% of the amount produced by solar cells, or a further 0.03% of global generation.

THE HISTORY OF SOLAR POWER

The exploitation of solar thermal energy can be traced back at least 29 centuries to the 7th century BC when the use of magnifying glasses was first recorded. This was followed in the 3rd century BC by the use of mirrors to concentrate the rays of the Sun, using the heat generated to light torches, often for religious ceremonies. The Greek scientist Archimedes is reputed, around 214–212 BC, to have used giant mirrors to focus light on the ships of Roman invaders who were attacking his home city of Syracuse in Sicily, apparently setting the wooden vessels on fire. There is no documented proof, but the experiment was recreated in the 1970s when a wood boat was set alight at 50 m. Meanwhile, throughout the last two millennia, architects and communities have used south-facing buildings to provide winter warmth.

A more systematic examination of solar energy was carried out by the Swiss scientist Horace de Saussure, who in the 18th century developed hot boxes in order to measure the heating effect of the Sun through glass. Hot boxes were later used for cooking.

The first successful attempt to convert solar energy into mechanical power was achieved by the French scientist Auguste Mouchout.[1] Mouchout used a reflector to concentrate the Sun's energy and generate steam in a glass-enclosed iron cauldron, and was able to use the steam to drive a small steam engine. With the support of the French government he continued to develop his device until 1881. Inspired by Mouchout's experiments, William Adams, an English civil servant in India, in the late 1870s developed the idea of using an array of flat mirrors arranged in a semi-circle to focus the Sun's energy on a single spot. This forerunner of the modern solar tower was able to drive a

---

[1]Much of the background for solar thermal power is based on the article: Revisiting Solar Power's Past, Charles Smith, Technology Review, Jul. 1995 (http://www.solarenergy.com/info_history.html).

2.5-horsepower steam engine during daylight hours. Around the same time in the United States, the Swedish-born engineer John Ericsson developed a parabolic trough reflector system.

Commercial exploitation of solar thermal power began with the Solar Motor Co., formed in Boston, United States, in 1900 by Aubrey Eneas. Eneas' company was not successful, but a company formed by another American entrepreneur, Frank Shuman, built the world's first successful solar-powered plant. Using parabolic troughs designed to track the Sun across the sky, the Sun Power Co. built a pumping station just outside Cairo, Egypt, that was able to generate 55 horsepower at a cost of $150/horsepower. The plant was completed in 1912 and appeared to be a success. Unfortunately the outbreak of World War I led to the destruction of the solar pumping station and the technology was not revived after the Armistice in 1918. However, in the 1960s parabolic reflectors were used in Italy to generate electric power, and commercial solar thermal power became a reality at the end of the 20th century.

The second means of generating electricity from the Sun, the solar cell, was born from the work of the French scientist Edmond Becquerel, who discovered the photoelectric effect. Becquerel observed that when light was shone onto the electrodes of a simple electrochemical cell made from two metal electrodes immersed in an electrolyte, the amount of electricity from the cell increased. This was followed in 1783 by the publication of a scientific paper in which an English electrical engineer, Willoughby Smith, revealed his observations that the electrical conductivity of the semiconducting material selenium increased when exposed to light. This was followed by further work on selenium by William Adams and Richard Day who, in 1876, discovered that illuminating a junction between selenium and platinum produced an electric current. This led to the production of the first solar cell made from selenium by the American inventor Charles Fritts. The conversion efficiency was around 1% and costs were too high for a practical device.

There was more work on the photovoltaic effect and on semiconductors during the next 60 years, including pioneering theoretical work by Einstein, but it was not until the early 1950s that true photovoltaic technology was born. This was the result of research by Daryl Chapin, Calvin Fuller, and Gerald Pearson at Bell Labs in the United States. The team produced a silicon photovoltaic cell with an efficiency of 4%, and later increased this to 11%.

Commercialization of solar cells was slow to take off, but in 1958 they were used to provide power to early satellites such as the U.S. Vanguard 1 and the Russian Sputnik-3. The technology remains the main source of electric power in space. Development continued throughout the next six decades, with new materials and higher efficiencies leading to cost-effective commercial solar cells for both central power and domestic rooftop applications; these became available by the middle of the second decade of the 21st century.

GLOBAL SOLAR POWER GENERATING CAPACITY

Records of solar cell capacity from the early years of their deployment are patchy, but the total global capacity appears to have been around 100 MW or less in 1990. In 1993 the installed capacity in IEA countries was 113 MW according to the IEA. Global capacity began to rise more rapidly during the 1990s, and by 2000 the total global capacity was 1288 MW, as shown in Table 1.1, while the global annual capacity addition was just over 200 MW.

Table 1.1 Annual Global Solar Photovoltaic Installed Capacity		
Year	Annual Capacity Addition (MW)	Total Global Installed Capacity (MW)
2000	293	1288
2001	324	1615
2002	454	2069
2003	566	2635
2004	1088	3723
2005	1389	5112
2006	1547	6660
2007	2524	9183
2008	6661	15,844
2009	7340	23,185
2010	17,151	40,336
2011	30,133	70,469
2012	31,011	100,504
2013	38,352	138,856
2014	40,134	178,391
Source: EPIA, SolarPower Europe.[2]		

[2]Global Market Outlook for Photovoltaics 2014–2018, European Photovoltaic Industry Association, 2014 and Global Market Outlook for Solar Power 2015–2019, SolarPower Europe, 2015.

The figures in Table 1.1 show both the aggregate annual capacity of solar cells and the annual capacity additions. The latter reflect the overall global manufacturing capacity. Total global aggregate capacity reached 5112 MW in 2005 and 40,336 MW in 2010, when over 17,000 MW of capacity was added during the year. Since then, annual capacity additions have been 30,000 MW or more, so that by the end of 2014 the total global installed capacity was estimated to be 178,391 MW. Market predictions put the capacity at the end of the second decade of the 21st century at between 400 and 600 GW.

While the use of solar cells is now widespread, the early growth in solar cell deployment was driven by national programs that provided incentives for rooftop installation of solar panels. The most notable of these were in Japan, Germany, and the U.S. state of California. Deployment in Germany spread to other European countries during the first decade of the 21st century, with a large rollout of cells in Spain, and then in Italy and France. All these contributions have led to Europe having the largest regional installed capacity of solar cells, 81,488 MW at the end of 2013, as shown in Table 1.2. However, in 2014 Europe only installed around 7 GW of new capacity, lower than the United States, Japan, or China. Of this, 2400 MW were installed in the United Kingdom and 1900 MW in Germany.[3]

The Asia Pacific region, including Japan, had the second largest regional installed capacity at the end of 2013 with 21,992 MW, while

Table 1.2 Regional Solar Photovoltaic Capacity and Capacity Additions, 2013		
Region	Capacity Addition, 2013 (MW)	Regional Capacity, 2013 (MW)
Europe	10,975	81,488
Asia Pacific	9833	21,992
China	11,800	18,600
The Americas	5362	13,327
Middle East and Africa	383	953
Rest of the world	n/a	2098
Total	38,532	138,856
Source: EPIA, SolarPower Europe.[4]		

[3]Global Market Outlook for Photovoltaics 2014–18, European Photovoltaic Industry Association, 2014; and Global Market Outlook for Solar Power 2015–19, Solar Power Europe, 2015.
[4]Global Market Outlook for Photovoltaics 2014–2018, European Photovoltaic Industry Association, 2014.

China had a further 18,600 MW. During 2014, China added 10,600 MW, pushing its total capacity to 29,200 MW. Meanwhile, an additional 9700 MW installation in Japan pushed the total in the Asia Pacific region to over 30,000 MW.

Total installed capacity in all the Americas at the end of 2013 was 13,327 MW. In 2014 6500 MW were installed in the United States, pushing the total across the Americas to above 20,000 MW. The lowest regional capacities at the end of 2013 were in the Middle East and Africa, with a combined aggregate of 953 MW. However, this increased substantially in 2014 when South Africa alone installed 800 MW. The solar photovoltaic capacity in the rest of the world at the end of 2013 was 2098 MW.

Table 1.3 contains figures for the installed solar thermal generating capacity in various regions of the world at the end of 2013, based on figures from the IEA. The total amount at the end of that year, 3500 MW, is much smaller than the solar photovoltaic capacity, reflecting the much slower market growth of this technology. Of this amount, around 1300 MW are fitted with some form of thermal energy storage. However, plants with energy storage are expected to become much more common by the end of the second decade of the 21st century.

The largest regional solar thermal capacity at the end of 2013 was based in Europe, virtually all of it in Spain. However, development there has slowed significantly. The capacity in the Americas at the end of 2013 was 900 MW, mainly concentrated in the United States. Here

Table 1.3 Regional Solar Thermal Generating Capacity, 2013	
Region	Installed Generating Capacity (MW)
The Americas	900
Europe	2300
Asia and Oceania	100
Africa	100
Middle East	100
Total	3500
Source: International Energy Agency.[5]	

[5]Medium Term Renewable Energy Market Report 2014, International Energy Agency 2014.

Table 1.4 Global Electricity Production From Solar Power Plants			
Year	Solar Photovoltaic Production (TWh)	Solar Thermal Production (TWh)	Total Solar Production (TWh)
2002	1.2	0.6	1.7
2009	20.0	0.9	21.0
2010	31.8	1.7	33.5
2011	60.8	2.3	63.1
2012	100.4	4.1	104.5
Source: Observ'ER.[6]			

a further 600 MW were added in early 2014, pushing aggregate capacity to 1500 MW. Capacity elsewhere is limited, with 100 MW in each of Asia and Oceania, the Middle East, and Africa. The IEA has predicted that cumulative capacity may reach 11,000 MW by 2020.

Total electricity production from solar power plants between 2002 and 2012 is shown in Table 1.4, based on data from the Fifteenth Inventory published by Observ'ER in 2013. The figures show that total solar production increased from 1.7 TWh in 2002 to 104.5 TWh in 2012. Most of this growth is accounted for by production from solar cells, which was responsible for 100.4 TWh of production in 2012. The contribution from solar thermal plants was 4.1 TWh. Global solar capacity nearly doubled between 2012 and 2015, and it is likely electricity production has nearly doubled too.

[6]Worldwide electricity production from renewable energy sources, Fifteenth inventory, 2013 edition.

CHAPTER 2

The Solar Resource

Solar energy is energy produced within the body of the Sun and then radiated into space. This energy from the Sun is generated during nuclear fusion reactions that take place inside the body of the star, within the core, a region that occupies about one-sixteenth of its volume. Within this region the temperature and pressure are high enough to promote fusion reactions between protons (hydrogen nucleii), which produce helium nucleii, at the same time releasing large quantities of energy. This emerges in the form of high-energy gamma radiation.

The temperature inside the core of the Sun is around 15,000,000 K, and the amount of energy produced is 3.86×10^{27} Joules/second, which corresponds to the conversion of 600 million tonnes of hydrogen each second. However, because of complex interactions, the gamma ray photons carrying this released energy can take over a million years to pass through the outer layers of the Sun toward the surface. By the time they reach the Sun's surface, the temperature has cooled to 6000 K, the temperature that can be registered by measurement from the Earth. The energy emitted from the surface of the Sun amounts to an average of close to 230 million W for each square meter. While this may appear extreme, the conditions found within the Sun are similar to many stars in our galaxy (the Sun is classified as a yellow dwarf).

The energy emitted from the surface of the Sun is mostly in the form of ultraviolet, visible, and infrared radiation, as well as massless, chargeless particles called neutrinos. The energy is radiated in all directions and its intensity diminishes as the distance from the Sun increases—that is, as the density of solar rays becomes smaller. At the distance of the Earth from the Sun, which averages 149,600,000 km, the amount of solar radiation passing through a square meter perpendicular to the direction of the Sun's rays is, according to the most recent estimates, 1361 W/m^2.[1] This figure is commonly known as the solar constant.

---

[1] The World Meteorological Organization puts the value at 1367 W/m^2.

Solar Power Generation. DOI: http://dx.doi.org/10.1016/B978-0-12-804004-1.00002-6

At the point at which the radiation from the Sun reaches the Earth, before it enters the Earth's atmosphere, the light waves fall mostly within the wavelength range of 200 nm (ultraviolet) to 2500 nm (infrared). At this stage its composition is roughly 56% infrared radiation, 36% visible radiation, and 7% ultraviolet. The remaining 1% is found mostly at longer wavelengths.

Not all of this radiation reaches the surface of the Earth, as illustrated in Fig. 2.1. Some of it will be scattered by dust and molecules within the atmosphere. Scattering of this sort is a random process, sending the radiation in all directions. Some will be scattered back into space while the remainder will fall toward the Earth's surface as diffuse radiation. More radiation is reflected back into space by clouds, and these play an important role in regulating the temperature of the atmosphere and of the Earth's surface.

Another part of solar radiation is absorbed by atoms and molecules in the atmosphere. Nitrogen and oxygen absorb short wavelength ultraviolet radiation, thereby blocking radiation with a wavelength shorter than 190 nm. Oxygen molecules can split into atoms by absorbing short wavelength radiation, leading to the production of ozone, which absorbs slightly higher wavelength ultraviolet radiation. Meanwhile, water vapor, carbon dioxide, and molecular oxygen also absorb in the

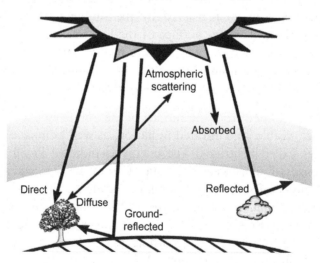

Figure 2.1 The components of solar radiation in the Earth's atmosphere. Source: Newport Corporation.[2]

[2]http://www.newport.com/Introduction-to-Solar-Radiation/411919/1033/content.aspx.

infrared region of the spectrum. When all these effects are taken into account, the flux of direct radiation that reaches the Earth's surface is roughly 1050 W/m² perpendicular to the direction of the radiation. When the additional scattered radiation reaching each square meter at ground level is added to this, the total flux is 1120 W under optimum conditions. By this stage the composition of the radiation is roughly 50% visible radiation and 47% infrared (Table 2.1).

Other factors will also modify the intensity of radiation reaching the surface of the Earth. One important factor is the rotation of the Earth. When the Sun is directly overhead at midday, the radiation has the least distance to pass through the atmosphere before it reaches any specific area of the ground. However, as the Earth rotates that same area on the ground away from midday, the distance through the atmosphere that the rays must pass increases. At the point at which the Sun appears to set from that point on the Earth, and no direct radiation reaches the surface there, the distance through the atmosphere that the last rays must travel is at its maximum. Beyond sunset there is no further direct radiation, but there is still scattered radiation that continues through evening until night falls and no sunlight remains.

All these factors reduce the amount of energy reaching the Earth's surface. As a consequence, the average solar flux on each square meter of the Earth—taking into account the fact that each square meter is in darkness for part of the time—is 170 W/m². The highest flux, found near the Red Sea in the Middle East, is 300 W/m². This is less than one quarter of the flux equivalent to the solar constant.

The effects of the atmosphere on solar intensity are quantified in terms of a factor called the Air Mass (AM) factor. AM0 corresponds to the intensity of the sunlight at the edge of the Earth's atmosphere.

Table 2.1 Solar Energy and the Earth	
Temperature inside the Sun's core	15,000,000 K
Energy produced by the Sun	3.86×10^{27} Joules/second
Solar constant	1361 W/m²
Composition of solar radiation reaching the atmosphere	56% Infrared radiation, 36% visible radiation, and 7% ultraviolet
Average amount of solar flux reaching the Earth's surface	170 W/m²
Total solar energy reaching the Earth each year	3,400,000 EJ

AM1 represents the intensity of sunlight at the surface of the Earth when the Sun is directly overhead, at the zenith, without any cloud cover. It therefore represents the attenuation caused by the column of air between a square meter on the Earth's surface and a square meter at the edge of the atmosphere. A standard intensity that is used when testing solar cells is AM1.5. This is the typical solar intensity found at midday in many of the main population centers of the Earth. It corresponds to 1000 W/m^2.

Even with all the attenuation and absorption effects discussed above, the quantity of solar energy reaching the Earth is enormous. The total solar flux reaching the disk of the Earth is 1.08×10^8 GW, while the total amount of energy that falls onto the Earth from the Sun each year is 3,400,000 EJ, or between 7000 and 8000 times global primary energy consumption; the solar energy falling on the Earth in one hour would supply current annual energy demand. Only a tiny proportion of this would be required to provide the 5000 GW to 6000 GW of electricity generation currently available across the globe. It is for this reason that solar energy is considered to be the best and most valuable source of renewable energy, and the resource most capable of displacing fossil fuels in a carbon-emission-free world.

As already noted, the solar radiation that reaches ground level on the Earth is of two types: direct radiation and diffuse radiation. The latter is produced when direct radiation is scattered and reflected as it passes through the atmosphere. Vegetation can absorb both types of radiation, allowing photosynthesis and plant growth to continue under all light conditions. Planar solar cells can also absorb both direct and diffuse solar radiation. However, solar thermal power plants and concentrating solar photovoltaic technologies both require direct solar radiation in order to operate effectively. This limits their application to regions where there is low average annual cloud cover. When there is no cloud cover, between 80% and 90% of the direct radiation passing through the atmosphere will reach the surface as direct radiation.

INTERMITTENCY

While the solar resource is vast, it is also intermittent. The Sun shines during the day but it does not do so at night. This diurnal variation means that solar input can usually only be relied on for part of the 24-hour daily cycle (see Fig. 2.2). In addition, the intensity of the solar

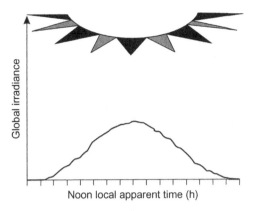

Figure 2.2 The diurnal variation in solar radiation on a sunny day. Source: Newport Corporation.[2]

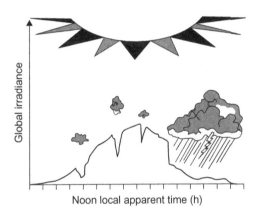

Figure 2.3 The diurnal variation in solar radiation on a cloudy day. Source: Newport Corporation.[2]

radiation falling at any given point will vary by the minute and by the hour depending upon the time of day and weather conditions, such as the amount of cloud cover, as shown in Fig. 2.3.

This means that on its own solar energy cannot provide a continuous source of power anywhere in the world. In order to make energy continuously available, either another source of energy must also be provided to back up the solar power, or a means of storing solar energy is needed. Both approaches are in use today.

While many grids have significant solar inputs, it is rare that the solar energy will be sufficient to supply all users on the grid, so there will normally be a range of other sources of electricity available, some renewable—such as wind energy—and others based on fossil fuels. A range of energy storage systems are also capable of storing solar energy when the Sun is not shining. There is limited use of these on grid systems today, but they are much more common in off-grid solar power systems where batteries are charged with solar energy during the day for use overnight. The development of cheap storage systems is expected to be one of the key technologies that will enable all renewable technologies—especially solar and wind power—to provide a large proportion of the global electricity supply in the future.

DISTRIBUTION OF SOLAR ENERGY

The orientation of the Earth with respect to the Sun means that solar energy that falls on the Earth is not evenly distributed. More falls on land and sea close to the equator, while less falls in the more northerly and southerly regions of the Earth, toward the poles. Global solar energy distribution maps show that the regions with the highest levels of solar irradiation are found across Africa and the Middle East into Iran, Afghanistan, and parts of India; in Australia; in the southern regions of North America, particularly Mexico and the southwestern United States; and in some parts of South America, including Brazil and the western coastal desert regions. This can be seen in Fig. 2.4, which shows

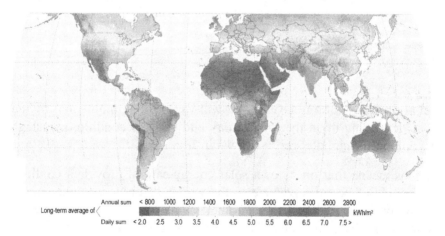

Figure 2.4 Global horizontal solar irradiation map. Source: Global horizontal irradiation map (SolarGIS © 2016 GeoModel Solar[3]).

[3]http://solargis.com.

the global long-term average horizontal radiation across the globe. Global horizontal radiation is the total amount of solar irradiation, both direct and scattered, that falls on a defined area at ground level. Fig. 2.5 shows a similar map of the global direct radiation.

The solar distribution affects planning and strategizing for a future in which fossil fuels are no longer widely used to generate electricity, and solar energy becomes a major resource replacement. In the United States, for example, the use of the southwestern region (which has high insolation) as a major source of solar electricity generation to supply the whole the country has been discussed. Europe has limited solar resources but close by, across the Mediterranean Sea, North Africa has vast quantities of solar energy. This energy could be captured using solar power plants, and then the electricity generated could be shipped through transmission lines across the Mediterranean Sea to Southern Europe. Proposals along these lines have also been discussed.

Other major energy-consuming regions will also be looking at solar potential. India has some regions with good solar resources, particularly in the northeast, and these could be exploited in an effort to replace coal, which is currently the main source of electricity in the country. China also has the potential to generate power from solar energy in the country's southwestern region, and in some more central regions too.

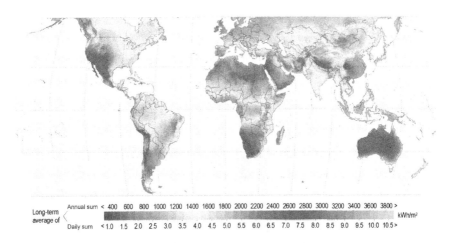

Figure 2.5 Global direct solar irradiation map. Source: Direct normal irradiation map (SolarGIS © 2016 GeoModel Solar[3]).

To put solar potential into perspective, consider a group of solar thermal power plants built in the U.S. state of California during the late 1980s and early 1990s. These power plants were pioneering solar plants, and used solar reflectors to capture the Sun's energy and the heat to drive a steam generator. The solar stations—which incidentally are still in operation in the second decade of the 21st century—were designed on the basis that the site could provide 2725 kWh/m^2/year, or 22.75 GWh/year for each hectare of land. Assuming that this incident energy could be converted with around 10% efficiency (modern solar plants hope to achieve better than this), then 2 million hectares or 20,000 km^2—a land area of roughly 150 km by 150 km—would be able to supply all the electricity for the entire United States.

CHAPTER 3

Solar Thermal Power Generation

Solar thermal power generation technologies exploit the Sun as a heat source. A typical solar thermal plant captures the infrared radiation that falls on the Earth and uses it to heat a thermodynamic fluid in order to drive a heat engine. A solar thermal facility is able to exploit a small amount of the long-wavelength visible radiation too, and power-generating plants are able to use around 50% of the energy in sunlight that reaches the surface of the Earth. However, these types of power plants can only exploit direct radiation. Diffuse radiation is of no use because it cannot provide the necessary intensity.

The heat engine in a solar thermal plant is the device that converts heat energy into mechanical energy; it drives a rotating shaft, which turns a generator to produce electricity. The heat engine extracts energy from a thermodynamic fluid by means of a cycle through which the fluid passes and via a temperature gradient. The larger this temperature gradient, the more efficient the engine is at extracting energy. To create the temperature gradient there must be a hot source to heat the fluid on one side of the cycle and a cold source to cool it on the opposite side. The cold source is usually either air or water, while the hot source for a thermal power plant is the Sun.

The radiation that arrives directly from the Sun, without any reflection or scattering, will feel hot, but its intensity is relatively low. As a consequence it cannot, as it arrives, provide a sufficiently intense hot source for a thermal power plant. The temperature rise that can be achieved is simply too low for the heat engine to operate efficiently.[1] In order to provide the intensity of radiation needed to heat a fluid to the temperature required to drive a heat engine efficiently, the solar radiation must be concentrated. This can be carried out using mirrors or lenses. For solar thermal power plants, mirrors are most commonly

---

[1]There is an exception to this, the thermal pond, which does exploit the Sun's infrared radiation as it falls onto the Earth, but these devices are relatively inefficient sources of electric power. They are discussed later in the book.

Solar Power Generation. DOI: http://dx.doi.org/10.1016/B978-0-12-804004-1.00003-8

used; the main types of solar thermal power plants are defined by the way in which they concentrate solar energy.

The solar collection field comprises one of the major sections of a solar thermal power plant. The second major section is the energy conversion system. Some solar thermal power plants may also include a third section, for energy storage.

GLOBAL SOLAR THERMAL POTENTIAL

As with solar energy potential in general, as discussed in Chapter 2, the potential for solar thermal generation is enormous. However, while solar photovoltaic power plants based on solar cells can be sited more or less anywhere, solar thermal power plants need a large area of land where there is good annual direct radiation and little cloud cover. This restricts their use but still leaves a vast exploitable resource.

Table 3.1 shows figures for this potential, by global region, for sites with high incident radiation intensity and limited land use restrictions. The aggregate global potential based on these figures is 12,150,000 TWh/year. This can be compared to total global electricity generation in 2013 of 23,322 TWh, according to the International Energy Agency (IEA).[2] On this basis, solar thermal power generation could provide 520 times the electricity currently being produced.

The regional breakdown in Table 3.1 shows that in North America there is the potential to generate 1,150,000 TWh each year from solar

Table 3.1 Regional Solar Thermal Potential	
Region	Potential Power Generation (1000 TWh/year)
North America	1150
South America	1350
Africa/Europe/Asia	7350
Pacific	2300
Total	12,150
Source: SolarPACES 2009.[3]	

[2]Key World Energy Statistics 2015, International Energy Agency.
[3]Global Energy Supply Potential of Concentrating Solar Power, Christian Breyer and Gerhard Knies, Proceedings SolarPACES 2009.

thermal power plants, and in South America, 1,350,000 TWh. The three regions—Africa, Europe, and Asia—are grouped together in the analysis on which these figures are based; the total potential for these three regions is 7,350,000 TWh/year. For the Pacific region the total annual potential is estimated to be 2,300,000 TWh. It is important to bear in mind that this regional potential is often found in desert areas far from population centers, so power generated in these regions might have to be transported over long distances. Even so, with modern long distance power transmission technologies such as high-voltage direct current (HVDC) transmission, it would be feasible for the total world electricity supply to be provided from solar thermal power plants.

SOLAR THERMAL TECHNOLOGIES

As noted above, typical solar thermal power plants comprise several elements. There will be a collector field that collects and concentrates the heat energy arriving from the Sun. The concentrated heat energy is then used to raise the temperature of a heat collection fluid that is positioned at the point where the energy is concentrated or focused. This hot fluid may drive a heat engine directly or it may itself be used to heat a second thermodynamic fluid such as water, converting it to steam to drive a steam turbine. Once the heat has been extracted from the first fluid, it is returned to the solar field to be reheated.

There are three primary configurations that have been developed for solar thermal power plants. They all have common features, but each is normally considered a distinct type of solar thermal plant. The three, defined by the way in which they collect and concentrate sunlight, are called parabolic trough (or solar trough) power plants, solar tower power plants, and solar dish power plants. Two other solar thermal technologies, called solar ponds and solar chimneys (or sometimes, solar updraft towers), have also been developed, though neither has yet been deployed commercially on any scale.

In a solar trough power plant the solar collection field is made up of rows of reflective troughs, each with a parabolic cross section. The parabolic reflector focuses the sunlight to a line along the length of the trough; it is here that a tube carrying the heat absorbing fluid is placed. These plants have proved to be some of the most cost-effective solar thermal generating plants, and there are now a large number of such power stations in operation.

The solar dish power plant uses a parabolic dish reflector instead of a trough. The circular reflector collects all the light within its coverage area, and focuses it to a single point where the heat absorbing element of the plant is situated. This central focus allows a higher intensity to be achieved than is possible with a trough, and a consequent higher temperature to be reached. However, the design is limited by the size of the dish reflector that can be constructed. Individual solar dish plants are relatively small. Large numbers of these are needed to make a high-capacity generating plant.

The solar tower is, in a sense, a way around the problem of constructing a massive solar dish. This type of plant has a tower with a heat collector mounted on top. This tower is then surrounded by a field of flat plate (or slightly parabolic) reflectors, all mounted at ground level and all oriented so that sunlight reaching them is reflected onto the heat collector on top of the tower. This reflector field approximates a parabolic dish. As with the solar dish, this arrangement generates a high solar intensity at the collector and potentially provides for a high efficiency heat engine.

One of the drawbacks of solar energy is the fact that it is only available during daylight hours and, in the case of a solar thermal plant, it can only be used to produce electricity when a high enough level of direct radiation is available. In order to get around this, solar thermal power plants can be equipped with some form of thermal heat storage. In a plant with heat storage, the solar energy collected during daylight hours is used to heat the storage medium. Heat is then extracted from this medium using a second heat transfer cycle, and this heat is used to drive the heat engine. If the size of the heat store is large enough, it can hold heat energy sufficient to provide power for a full 24 hours each day.

UNCONVENTIONAL SOLAR THERMAL TECHNOLOGIES

In addition to the three solar thermal generating plant configurations outlined above, there are two others that have either been proposed or tested: solar ponds and solar chimneys. The solar pond is a very low-technology approach to solar thermal power generation. It consists of a large pond filled with brine, which is heated by the Sun. As the pond warms, a temperature gradient develops between the upper and lower layers of the pond, and this temperature gradient is used to drive a small heat engine. The temperature gradient is never very large, and so the system depends on a low-temperature heat engine of special design, a type often used to extract energy from low-temperature geothermal resources.

The second unconventional approach is completely different. It uses a large greenhouse to heat air, which is then funneled towards a tall tower, thus creating a powerful updraft. The mass of moving air generated by this draft is used to drive wind turbines, which are placed in the path of the flowing air. Plants of this type have to be massive to become economically cost-effective, so while the concept has been proved at a pilot scale, no plant of commercial size has ever been built.

SOLAR COLLECTORS

The majority of solar thermal power plants use mirrors to collect and focus solar energy. In most cases these mirrors are made from silvered glass. Glass is extremely durable and can be molded into the shapes necessary for the solar reflectors. Flat plate glass is cheap, but creating the parabolic shapes needed for solar reflectors adds to the cost. In addition, glass can be heavy, particularly with large area reflectors that must be physically strong enough to withstand any bending forces resulting from their own weight.

Alternatives to glass have been used in solar plants, but their use is relatively uncommon. Plastics such as silver-coated acrylic have been tested, and so has polished aluminum. Novel techniques for creating the required shapes, such as using a vacuum to shape a circular metal sheet or using arrays of flat sections to simulate a parabola, have also been used. However, glass has generally proved to be the most reliable and durable material for solar collectors.

Once the solar heat has been collected it must be absorbed by a heat transfer medium. This is carried out in a variety of ways. In solar trough plants a fluid is pumped through pipes that run along the focus of each trough. This fluid may be a high-temperature oil, or it can be water and steam. In other solar tower power plants a mixture of molten salts, or liquid sodium, have been utilized. It is also possible to use a pressurized gas. In some cases the heat transfer fluid is used directly in the heat engine of the plant, but often there is a secondary heat exchanger that transfers the heat from the transfer medium to the thermodynamic fluid. Both systems have their advantages.

Low-technology solar ponds do not use any form of concentration. Direct sunlight heats the pond without any additional measures. This means that only a modest temperature rise is possible, and efficiency is relatively low.

ENERGY STORAGE

Energy storage offers a means of extending the range of solar thermal power plants beyond the hours when heat from the Sun can be collected. Most solar thermal plants have a limited thermal inertia as a result of the volume of heat transfer fluid they contain (and its thermal mass), but this will not allow the plant to operate for more than a few minutes without heat input.

It is possible to build additional thermal storage capacity into plants in a number of ways. In solar tower power plants a common approach is to use a molten salt as the heat transfer and storage medium. Heat can be stored in large vessels full of the molten salt from which it is extracted by passing the salt through a heat exchanger to heat a thermodynamic fluid. Massive solid structures made from concrete, glass, or ceramics can also act as heat stores; these might provide high thermal storage density in the future. The solid/liquid or solid/gas phase change can also be exploited by using the latent heat needed to effect the phase change to store heat energy.

Some solar thermal power plants use a water/steam cycle as both the heat transfer and the thermodynamic fluid. In these plants it is possible to store energy in steam accumulators. This is not as efficient as some of the other methods available, but it has the advantage of simplicity and ease of interfacing with the plant.

According to the IEA, thermal storage is more efficient than the more common electrical energy storage technologies such as pumped storage hydropower.[4] This suggests that further development of heat storage technologies would be economically effective. One way of improving efficiency is to raise the temperature at which heat is stored.

[4]Technology Roadmap: Concentrating Solar Power, International Energy Agency, 2010.

However, this depends on being able to achieve higher temperatures when concentrating solar energy.

ENERGY CONVERSION AND HEAT ENGINES

Solar thermal power plants have used a range of heat engines to convert solar energy into electricity. For the low-technology solar pond, the most useful heat engine is probably an organic Rankine cycle engine. This can operate using a very small temperature difference between the hot and the cold source. The engine uses a low boiling point liquid so that the low temperature can be exploited, but in other respects the cycle is similar to that of a steam turbine.

Steam turbines are the most common means of extracting energy from a solar thermal plant. Some solar stations heat water directly to generate steam, but many use an intermediate heat transfer fluid—often an oil—which then heats the water through a heat exchanger. Direct systems are potentially more efficient, but they have proved to be more difficult to design. Meanwhile, the use of a steam cycle means that it is simple to add some form of supplementary heating, such as the addition of gas burners, to enable the plant to operate when solar heat is not available or is limited.

It is also possible to use the Brayton cycle gas turbine in a solar thermal plant. In this case the heat transfer medium must be a gas that can be heated to a high temperature and pressure. This can be achieved in solar tower plants and in solar dish plants where small gas turbines have been used as the heat engines. Solar dishes have also used another, novel engine, the Stirling engine. This is a closed-cycle engine with heat applied to its outside.

With any heat engine cycle it is important to have a cold source to remove heat from the engine. This is typically provided by air or water cooling. Many solar thermal power plants operate in arid desert regions where water is scarce, so they usually need to be able to use air cooling.

ELECTRICITY TRANSPORTATION

Solar thermal generating potential is often concentrated in regions of high insolation, such as the southwestern United States or North Africa. These regions, often arid or desert, can in principle provide large volumes of electricity from the Sun. However, they are also often remote from the main population and demand centers, and so dedicated power transmission systems are likely to be needed if such resources are extensively exploited.

If solar thermal generating regions are developed in the future, then the power is likely to be transported using long HVDC transmission networks. For long-distance transmission, this type of system has lower energy losses than a conventional alternating current (AC) system. Most HVDC lines are point-to-point lines that typically carry power from a large power plant or agglomeration of power plants to a single converter station in a distant AC grid. These lines may be more flexible in the future, exploiting new technologies now being developed that provide the ability to build branching HVDC networks.

A system of this sort has been proposed to carry power from the southwestern states of the U.S. to the population centers of the northeast. A similar type of scheme has been suggested to transport power generated in North Africa across the Mediterranean Sea to provide energy for European countries. So far these projects remain no more than proposals, but they do offer a possible future means of mitigating global warming.

CHAPTER 4

Parabolic Trough and Fresnel Reflector Solar Power Plants

The parabolic trough solar power plant takes its name from the trough-shaped reflector that is used to capture and concentrate the solar heat energy. The collector is much longer than it is wide, and has a parabolic cross section. Each trough is aligned on a north–south axis and provided with a system to allow it to track the Sun across the sky. As it follows the movement of the Sun, the reflector focuses the sunlight onto a solar energy receiver that runs along the length of the trough and is positioned at its focus. These concentrators are sometimes called line focusing solar thermal plants because the sunlight is concentrated along a line.

Closely related to the parabolic trough is the Fresnel reflector, a simplified version of the parabolic trough in which the trough shape is approximated by a series of long, flat—or nearly flat—reflectors that are generally mounted on the ground. The solar energy receiver is mounted separately from the reflectors on a framework that places it above the reflectors. This system is not as efficient at concentrating solar energy as the solar trough, but it is significantly cheaper to construct. The aim of the Fresnel design is to achieve simplicity and low capital cost. However, the technology is less well tested than the more conventional solar trough system.

LINE FOCUSING ORIGINS

The parabolic trough is the oldest of the modern solar thermal technologies. The first recorded version is that of a Dr. Maier of Aalen and a Mr. Remshalden of Stuttgart, who developed a system based on a parabolic trough collector to generate steam. Their system was followed in 1912–13 by a facility built in Cairo by U.S. inventor Frank Shuman. The plant used tracking solar troughs to generate steam for a

Solar Power Generation. DOI: http://dx.doi.org/10.1016/B978-0-12-804004-1.00004-X

steam engine, although the initial plan was for the plant to generate electricity.[1] The project comprised five collectors, each 62 m long and 4 m in width. The steam that the collectors were able to produce was equivalent to a generating capacity of 41 kW.

The next time the technology appeared in commercial form was in the 1980s in California. A company called Luz built nine plants based on solar parabolic troughs between 1984 and 1991. The first of these had a generating capacity of 13.8 MW, and the final plants had capacities of 80 MW. The technology was considered economically marginal when the plants were built, and in 1991 the builder filed for bankruptcy, unable to secure the financing to build a tenth plant. In spite of that, the nine plants continue to operate and generate power as of 2016. No more plants of this type were constructed until 2007, when the technology enjoyed a renaissance. Since then over 30 power plants based on this technology have been built, and it is arguably the most successful solar thermal technology today.

The Fresnel reflector system is based on a similar principle to the Fresnel lens that was originally developed for lighthouses by the French physicist Augustin-Jean Fresnel. The earliest example of a Fresnel reflector for concentrating solar power was developed by an Italian, Giovanni Francia, who patented his work in 1962. A prototype based on his design was built in France in 1964. The technology did not thrive then, but it was picked up again in Australia in the 1990s, where a large-scale demonstration project was built. Since then plants have also been constructed in the United States and in Spain.

PARABOLIC TROUGH TECHNOLOGY

A parabolic trough is a special type of solar concentrator that has a parabolic cross section (it is parabolic in two dimensions) but is linear in the third dimension. The result is that the parabolic shape is extended linearly to make a long reflector. The shape of the reflector causes sunlight to be concentrated along a line at the focus of the parabola, a line that runs along the length of the trough. A heat receiver, normally a specially constructed pipe, is positioned exactly at this focus so that it can absorb the heat from the Sun. A heat transfer fluid is pumped

[1]Details of the plant at Al Meadi, Cairo, were published in the Electrical Experimenter Magazine in Mar. 1916.

through the pipe and carries the heat away. In most plants this fluid passes through a heat exchanger where it heats water to steam; the steam is used to drive a steam turbine generator. A schematic of a parabolic dish power plant is shown in Fig. 4.1.

A single parabolic trough may be up to 150 m in length. However, each trough is normally made up of sections that are generally 5−20 m in length. Each unit may be 5−8 m wide, with more recent plants aiming for greater width. The width is referred to as the aperture of the trough. Trough collectors are around 2 m deep. The modules are assembled together on some form of support structure, which in modern designs is often a space frame constructed from tubular metal sections. Each module must be carefully aligned with its neighbor on the space frame support, and the whole trough must be able to rotate about an axis along its length to track the Sun across the sky. Alignment of the troughs is north−south so that they can track east to west by rotation of the trough.

The traditional means of constructing the modules has been to use glass. The glass is hot-formed in bending plants and the precise shape is achieved using parabolic molds. The back surface is then silvered. Mirrors of this type can achieve 94−98% reflectance. However, this is a relatively expensive construction technique, even with modern technology, and alternatives are being sought. One method that has shown promise uses a polymer sandwich construction with a thin layer of silver between polymer layers to form the reflective surface. These polymer

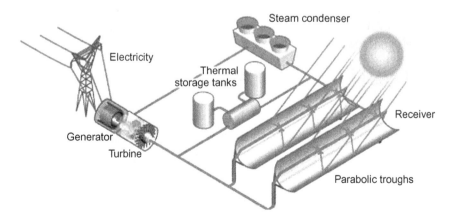

Figure 4.1 A parabolic dish power plant. Source: U.S. Department of Energy.

film reflectors are much lighter than glass and the mirrors are much less expensive to fabricate. A typical commercial product is around 60% lighter than the equivalent glass component, and 30–40% less expensive to fabricate.

Another alternative is to construct the reflector from a lightweight metal alloy, or from an acrylic structure with an alloy reflector. Both polymer film and alloy reflectors offer potential savings, but their long-term durability has yet to be proved. Research by the German Aerospace Centre has also suggested that they cannot achieve the same level of reflectivity as glass. With glass mirrors, the collector field is the most expensive part of a parabolic trough plant and can account for up to 50% of the total capital cost.

Parabolic trough reflectors can achieve a solar concentration ratio of between 60 and 100. The solar heat intensity this creates can potentially raise the temperature of a heat transfer medium to around 550°C. In order to achieve this temperature, the heat transfer fluid must be pumped through circuits that pass along several solar troughs. Many such loops will then be connected in parallel to provide the heat to raise steam. The actual temperature reached by the heat transfer fluid depends upon the fluid itself. Many plants use a synthetic oil that must not be heated above 400°C or it will start to decompose. For plants of this type, the temperature at the solar field outlet will typically be maintained at or below 390°C and the inlet temperature at 290°C so that the temperature rises by 100°C during the passage through the solar field.

A major alternative to the use of oil is a direct steam system in which the heat transfer fluid is a water/steam mixture. The advantage is that steam is generated within the solar field without the need for an intermediary heat exchanger, which significantly increases the overall efficiency. In addition, the temperature within the collection pipes is not limited to 400°C, which allows the power plant's steam turbine to operate at a higher efficiency. One drawback, however, is that a direct steam system creates many engineering challenges due to both the higher temperature in the collector field and the need to manage a water/steam system under changing heat input conditions. Consequently, direct steam production is not commonly used in commercial solar trough power plants.

A key component of the solar plant is the absorber tube that captures the heat needed to raise the temperature of the heat transfer fluid. This is typically a stainless steel tube that has been coated to increase its absorption properties. A typical receiver will be around 70 mm in diameter and built in 4–5 m sections. The receiver tubing may also have fins to aid absorption. The whole unit is then enclosed in a glass envelope, typically 110–120 mm in diameter, and the space between the steel tube and the glass outer envelope is evacuated to reduce heat loss. The steel absorber is mounted into the glass tube using a system of bellows that allows for relative expansion and contraction of the glass and steel. Junctions at the end of each collector must allow for the rotation of the receiver as the collectors track the Sun.

The efficiency of the energy collection system depends on the accuracy of its geometry as well as the efficiency of the heat absorbers. Solar heat capture efficiency can be as high as 75%.

FRESNEL REFLECTORS

The Fresnel reflector system simplifies the parabolic trough design. Instead of using single parabolic reflectors, it mimics the parabolic shape with a set of flat (or almost flat) mirrors mounted at ground level. A schematic of this type of plant is shown in Fig. 4.2. Typical Fresnel systems use 10–20 individual, long, reflecting segments instead of a single trough collector. Like a parabolic trough, these long, flat mirrors can be rotated about their long axes, which are oriented north–south, so that they can track the Sun across the sky. The use of

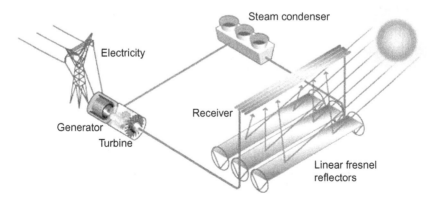

Figure 4.2 A Fresnel concentrator power plant. Source: U.S. Department of Energy.

flat mirrors instead of parabolic mirrors substantially reduces the cost of the collector field. In addition, the collection of mirrors focusing sunlight onto a single receiver can be much larger than is possible with a parabolic trough. In principle this also reduces costs because it can reduce the number of loops required for a similar-sized collector field. An additional advantage is that ground-level mounting of the mirrors reduces wind resistance, which can be a problem in desert regions where solar energy is most widely available.

In addition to the simplification resulting from segmenting of the parabolic reflector, the Fresnel system simplifies the heat-receiving system too, by using a set of absorber tubes fixed relative to the mirrors. Since they do not have to move, as would be necessary in the parabolic trough system, construction of the heat receiver circuit is simpler and therefore cheaper.

These simplifications mean that the Fresnel system is not as efficient at collecting and concentrating solar heat as a parabolic trough. A typical plant will achieve a concentration ratio of 30–50, only half that of a parabolic trough plant. Solar collection efficiency is lower too, with a maximum of 70%. On the other hand, these systems can be more compact at around 1–2 ha/MW. In order to increase efficiency, Fresnel reflector plants often use direct steam heat collection systems instead of the oil heat transfer fluid and intermediate heat exchanger common in parabolic trough plants, an option made possible by the simpler heat circuit in this type of plant.

POWER GENERATION

Solar trough and Fresnel reflector power plants generally use steam turbine generators to produce electricity. The heat transfer fluid exiting the solar field is passed through a system of heat exchangers that heat water and convert it into high-pressure steam. Plants often use reheat steam turbine cycles in which the steam turbine is split into two components. The hottest steam exiting the heat exchanger is fed into the first section of the steam turbine. The steam that exits this turbine is then heated again in a unit called a reheater before entering the second section of the turbine, increasing overall efficiency. Steam exiting the second turbine section is condensed, using a water or air-cooled condenser. The steam cycle water exiting the condenser then returns to the heat exchanger where it is preheated using the last of the heat from the heat transfer fluid before the latter is returned to the collector field.

The steam cycle, with a maximum temperature of around 400°C, can reach an efficiency of 35−38%. This yields an overall plant efficiency of 15−16%.

It is possible to boost the output of a solar thermal power plant such as a parabolic trough plant by using supplementary heating. This involves adding some form of gas firing to the heat exchanger. The additional heat input can be used to smooth out variations in solar input during the day and to extend the operation of the plant during the morning and evening hours.

ENERGY STORAGE

The amount of time that a solar thermal power plant can actually supply energy to the grid is limited by the fact that solar energy is only available during daylight hours. To circumvent this problem some solar trough power plants include heat energy storage systems. If some of the solar heat is stored during the day it can be used when solar input is no longer available, extending the period during which the solar plant can supply power.

The most common form of heat energy storage used in solar trough power plants involves storing heat in a mixture of nitrates, typically sodium nitrate and potassium nitrate. These materials are solid at room temperature but liquefy between 130°C and 230°C. The liquid can then be heated to between 500°C and 600°C before the nitrates start to decompose. This means that they are well matched to the solar thermal collector field where the heat transfer fluid temperature varies between 300°C and 400°C. A schematic of a plant of this type is shown in Fig. 4.3.

Heat storage is implemented by having a supplementary heat transfer fluid circuit that bypasses the steam generator heat exchanger. Fluid in this part of the circuit passes through a separate heat exchanger where the molten nitrate salt heat storage medium is heated. The storage system itself consists of two tanks, one with cool molten salt and one with hot salt. During energy storage, heat from the collector circuit is used to heat salt from the cold store, which is then passed

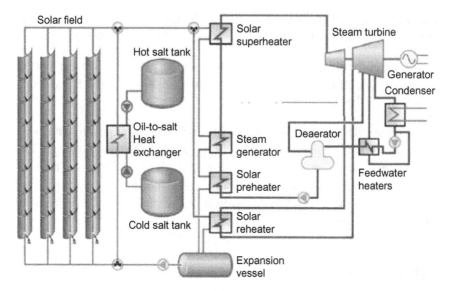

Figure 4.3 A parabolic trough power plant with heat energy storage. Source: Abengoa Solar – Innovative technology solutions for sustainability, A New Generation of Parabolic Trough Technology, SunShot CSP Program Review 2013, Phoenix, April 2013.[2]

into the hot store. When the heat energy is required, this process is reversed and the hot molten salt is used to heat the transfer fluid, which carries the heat to the steam generator section of the circuit.

The amount of energy storage that a plant can maintain depends on the size of the collector field. The larger the collector field in relation to the size of the steam turbine, the more heat available for storage. In theory it is possible to size the collector field to store sufficient energy to keep the plant operating 24 hours each day. In practice this is not the most economical design because the collector field is an expensive part of the plant. More typically a plant has energy stored for up to six hours of operation, which extends its output into the late evening.

COMMERCIAL SOLAR TROUGH POWER PLANTS

Since 2007 when interest in solar thermal technology resumed, around 30 commercial solar trough plants have been built. The majority of these plants are either in Spain or the United States, with a smaller number in the Middle East. The largest of these are two 280 MW

[2]http://www1.eere.energy.gov/solar/sunshot/pdfs/csp_review_meeting_042513_price.pdf.

facilities in the United States. One of these, in Arizona, covers an area of 780 ha, or 2.8 ha/MW, an area/MW ratio typical of these plants. The collector field consists of 3232 individual troughs, each made up of 10 modules. There are 808 collector field loops, with each loop passing through eight troughs. The solar field inlet temperature is 293°C and the exit temperature is 393°C. In addition, the plant has a molten salt energy storage system that uses two tanks to provide six hours of operation at full power. Total cost of the plant is estimated to be $2 billon.[3]

A much smaller number of Fresnel reflector power plants have been constructed. The largest is a 100 MW plant in Rajasthan, India. Others are below 50 MW.

HYBRID SOLAR THERMAL POWER PLANTS

Adding supplementary heating to a concentrating solar thermal power plant creates a form of hybrid plant. There is another configuration that involves adding a solar collection field to a traditional natural gas-fired combined cycle power plant. In this case heat from the solar collection field is used to supplement the heat input for the combined cycle power plant. These plants are generally called Integrated Solar Combined Cycle (ISCC) plants. A diagram of a typical ISCC plant is shown in Fig. 4.4.

As long as the amount of heat input is kept relatively low compared to the natural gas heat input, the solar efficiency will be much higher than can be achieved in a stand-alone solar trough power station; also, the energy efficiency of the combined cycle power plant is improved. The best balance is achieved when solar input is no more than 10%. If it rises higher, then the effect on the efficiency of the combined cycle plant when solar input falls off at night becomes significant. A low solar input also allows the combined cycle plant to be maintained on standby and kept at operating temperature without burning gas, which increases its flexibility and availability when it is being used for grid support.

---

[3]Plant statistics are from the U.S. National Renewable Energy Laboratory's database of concentrating solar power plants.

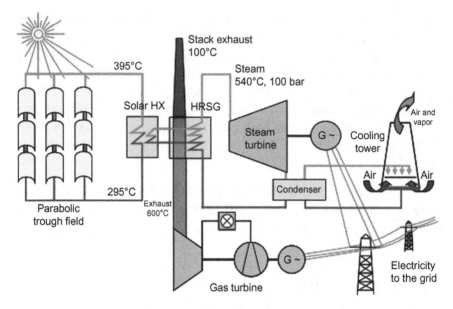

Figure 4.4 An integrated solar combined cycle plant. Source: Courtesy of SolEngCo GmbH – Georg Brakmann.

A small number of ISCC plants have been built over the past decade. They include projects in the United States, Australia, the Middle East, and Mexico. Solar fields in these plants are generally relatively small, typically under 100 MW, and often less than 50 MW.

CHAPTER 5

Solar Towers

A solar tower power plant is characterized by the way in which solar energy is collected and concentrated. This type of plant has at its center a tall tower carrying a thermal receiver. The tower is surrounded by a field of square or rectangular mirrors, usually called heliostats. The heliostats are mounted on special structures so that each can individually track the Sun across the sky and direct the sunlight onto the receiver at the top of the tower. Because all the solar energy is concentrated at one point instead of along a line as in a parabolic trough system, the level of concentration is higher in the former than in the latter. Solar towers are sometimes referred to as point focusing solar thermal plants.

The solar tower concept was first tested in Spain in 1981 at a 500 kW test facility that used liquid sodium as the heat transfer fluid to carry energy from the solar receiver to a heat exchanger to raise steam for a steam turbine. Soon afterwards, in 1982, a plant called Solar One was built in California with a water/steam heat transfer medium. The pilot plant, which was supported by the U.S. Department of Energy, had 1818 mirrors, each 40 m^2, and was able to generate 10 MW. The plant operated until 1986, but the technology was not developed commercially. Then, in 1996, Solar One was updated to a molten salt heat transfer system, and a ring of 108 mirrors of 50 m^2 was added. The new pilot plant, now called Solar Two, also had a generating capacity of 10 MW. Other pilot schemes were built in France, Italy, Japan, and Russia; the largest of these had a generating capacity of 5 MW.

These projects proved the technical feasibility of a large solar tower plant, but it was not until 2007 that the first commercial solar tower plant, called PS10, was built in Spain. The facility has a generating capacity of 10 MW. Since then there have been a small number of other commercial plants, including one in California comprising three central towers with a total gross generating capacity of 392 MW. A new plant based on the technology used in Solar One and Solar Two has also

Solar Power Generation. DOI: http://dx.doi.org/10.1016/B978-0-12-804004-1.00005-1

been constructed. This plant has the ability to store thermal energy; energy storage is relatively simple to integrate into some solar tower plants.

SOLAR TOWER TECHNOLOGY

The principle of the solar tower is the same as that of the solar trough: focus sunlight onto a solar receiver where a heat transfer fluid can be heated, and the heat carried away to generate electricity. With the solar tower the linear receiver is replaced with a single-point receiver mounted at the top of the central tower.[1] This receiver must be able to capture all the heat energy from a large number of heliostats mounted at ground level around it. Fig. 5.1 shows the layout of a typical plant of this type.

The type of receiver used in commercial solar tower power plants is called an external tube receiver. The solar heat directly hits the outside of tubes that carry the heat transfer fluid, and the heat is conducted through the tube material to the fluid inside. The efficiency of the receiver is crucial to the overall energy capture efficiency, so this represents a key

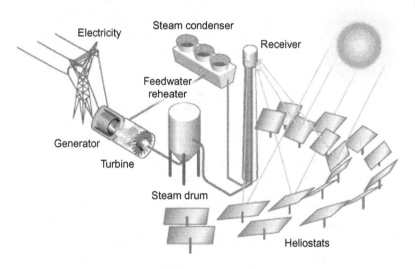

Figure 5.1 A solar tower power plant. Source: U.S. Department of Energy.

[1]Some recent solar tower power plants involve multiple towers.

area of research. New designs including a type called a volumetric receiver are being tested, and may offer better performance (see discussion of air-driven systems later in this chapter).

The heat transfer fluid that passes through the receiver varies depending upon plant design. Sodium, as used in the early Spanish pilot scheme noted earlier, has not been used in any commercial plants. However, some plants use molten salt as the heat transfer fluid, and this permits an element of storage to be included. Others use a direct steam system. The main advantage of the solar tower over the line focusing type of solar thermal plant is that point focusing allows a much higher temperature to be achieved in the receiver; this enables the thermodynamic heat engine attached to the system to operate at higher efficiency.

The collection system for a solar tower is the field of individual helio-stats, each angled to direct sunlight onto the central receiver. Heliostats vary in size from $10\,m^2$ to $120\,m^2$. The 10 MW PS10 solar tower in Spain has a field of 624 heliostats, or roughly 60 for each megawatt of generating capacity, and the collector field covers an area of 60 ha, or 5.5 ha/MW. Each heliostat is $120\,m^2$. This large size can make them difficult to align in high winds. Meanwhile, the Ivanpah solar tower project in California has 347,000 mirrors, each $7\,m^2$—with two mirrors to each heliostat—for its three solar towers, and a net generating capacity of 377 MW, or 460 heliostats/MW. The collection field for the three towers that make up this power plant cover an area of around 1420 ha, or 3.75 ha/MW. The facility also has the ability to supplement its solar input by burning natural gas in order to maintain its power output.

The collector array for a solar tower bears some resemblance to the Fresnel reflector. The ideal shape for the field would be a giant parabola with the receiver at the focus. Such a large parabolic reflector would be impossible to construct, so the solar tower breaks this down into a massive array of small mirrors mounted at ground level. In addition, although in theory each mirror should have a parabolic curvature, in practice this is usually approximated with flat mirrors. The terrain upon which the power plant is built does not have to be flat because elevation changes over the collector field have little effect on efficiency, an advantage compared to solar trough power plants. However, there are economic limits to plant size because once the collector field becomes too large, additional mirrors at the periphery contribute relatively less energy. The practical limit is likely to be between 200 MW and 400 MW for an individual tower.

A solar tower point focusing heat collection system can potentially achieve a concentration ratio of 600–1000, ten times that of a line concentrating system. This high level of concentration would allow the system to potentially provide a temperature of 1000°C and above, which would in principle allow the heat engine used in a solar tower power plant to operate with 20% more efficiency than in a solar trough plant. In practice, however, commercial plants have not yet achieved such high temperatures.

POWER GENERATION

The high thermodynamic cycle temperature that can potentially be achieved with a solar tower power plant means that a greater variety of different configurations are possible for extracting power. As already noted, a number of different heat transfer fluids have been tested, including liquid sodium—a material that has also been used in nuclear fast reactors—molten salts, and a direct water steam system. Experimental systems have also included the use of air or a gas as the heat transfer fluid; hybrid cycles involving a gas turbine are also being explored.

The role of the heat transfer fluid is to capture heat from sunlight within the receiver, and then carry this heat away so that it can be used to drive a thermodynamic heat engine. This can be achieved in one of two ways. The most efficient is for the heat transfer fluid to be the same fluid that is used in the heat engine. The alternative is to pass the heat transfer fluid through a heat exchanger where it heats the thermodynamic fluid that drives the heat engine.

Using the heat transfer fluid directly in the heat engine is the simplest system, and in principle the most efficient because it does not involve a heat exchanger. This is the basis for a direct water/steam system in which the heat in the solar receiver is used to generate steam directly; this steam is then used to drive a steam turbine. This type of system has been used in the Ivanpah solar tower project in California, where a steam temperature of 545°C is achieved at a pressure of 170 bar to drive the towers' three 125 MW steam turbines. Overall plant efficiency is around 18%. A direct steam system is also used in two Spanish plants, PS10 and PS20. These operate at relatively mild steam conditions of 250°C and 40 bar. The two Spanish plants also incorporate an element of energy storage in the form of steam storage tanks, but the amount of storage is very limited and the plants can effectively only operate during daylight hours. Overall plant efficiency is 15.5%.

With direct heat transfer systems using a water/steam cycle it is possible to add supplementary heating as a means of increasing the power output of the system when solar input is reduced. Supplementary heating is used at the Ivanpah plant to heat the steam system in the morning so that it can start operating more quickly when the Sun rises.

The other system that has been employed in commercial solar tower plants involves the use of a molten salt as the heat transfer fluid. The molten salt is usually a mixture of nitrates that form a liquid at the temperatures achieved in the solar tower plant. Typically the mixture includes 60% sodium nitrate and 40% potassium nitrate. This system was tested in the Solar Two project in California and has more recently been used in two commercial plants, one called Gemasolar in Spain and the second called the Crescent Dunes solar tower in the U.S. state of Nevada.

In this type of system the heat in the solar receiver is used to heat the molten salt to around 550°C. This hot salt is then pumped to a hot storage tank where it can be stored. In another circuit, hot molten salt is pumped through a heat exchanger where it heats water to steam. The spent salt is then pumped into a cold storage tank. The temperature here is around 300°C so that the salt always remains in the liquid state (it solidifies at around 220°C). The steam generated in the heat exchanger is used to drive a steam turbine.

Depending on the size of the solar field, the size of the molten salt storage system, and the size of the steam generator, this type of plant is capable of storing a significant amount of heat energy. The Gemasolar plant has a tower 140 m high surrounded by 2650 heliostats, each 120 m^2. Its molten salt system is sized so that it can generate power for 15 hours without any solar input, and this allows it to operate 24 hours a day. Overall plant efficiency is 14%. The Crescent Dunes project is also capable of operating for 24 hours each day.

AIR-DRIVEN SYSTEMS

There is an alternative approach to power generation in a solar tower that uses air as the heat transfer fluid. This has the potential to be more efficient than the liquid systems described earlier, but the cycle is more complex.

The heart of this type of thermodynamic cycle is a device called a volumetric air receiver. This type of receiver is constructed of a porous material that can be a wire mesh structure or porous ceramic elements. These can absorb heat throughout the volume of the device and transfer it to a fluid passing through the receiver. The receiver provides a much larger surface area for heat transfer than an external tube receiver, and can potentially offer greater efficiency. In an air system, air is pumped through the volumetric receiver where it is heated by the hot receiver elements. The hot air is then pumped away and used to raise steam in a heat recovery boiler, steam that will drive a steam turbine. In principle it should be possible to achieve air temperatures of up to 1000°C in this type of receiver, offering higher thermal-to-electrical conversion efficiency than with current commercial systems.

A more advanced air cycle is possible if the whole air system is pressurized. The pressurized air is heated in the volumetric receiver as before, perhaps reaching 1100°C in an ideal design. The hot, high pressure air can then be used to drive a small gas turbine. Upon exiting the turbine, the hot air is once again used to raise steam in a heat recovery boiler, and the steam is used to drive a steam turbine. The result is a type of combined cycle solar power plant. It is possible to add both heat energy storage and supplementary heating to such advanced pressurized systems. Volumetric receivers and the various associated power generation configurations are being tested at the pilot stage, but are not yet available commercially.

CHAPTER 6

Solar Dishes

A parabolic dish solar power plant is based on a single parabolic reflector, similar in shape to a satellite antenna. This reflector focuses light to a point where a heat engine is placed. The heat engine uses the concentrated solar heat to produce electricity. In concept, the solar dish is similar to the solar tower, but on a much smaller scale. The small scale allows a full parabolic reflector to be built, but limits the maximum size of the solar dish that can be constructed. Economic and physical constraints mean that most solar dish units have a generating capacity of 25 kW or less, although some larger individual systems have been built.

A single solar dish unit can be used for distributed power applications, providing power to a single domestic dwelling or small commercial operation. For larger power plants, multiple solar dishes can be installed in arrays. In principle this can provide tens or even hundreds of megawatts of generating capacity from a single power plant, although the largest plant of this type that has been constructed had a capacity of only 4.9 MW.

The advantage of the solar dish over other types of solar thermal power plants is its efficiency. A unit of this type can use solar energy to raise the temperature of the thermodynamic fluid in a heat engine to 1000°C; solar dish power units have demonstrated an efficiency of 30% and higher. Another advantage is that the dishes can be self-contained generating units that do not require water cooling, which is often needed in other types of solar thermal power plants.

Most solar dish designs use simple, self-contained heat engines, one to each dish. However, some parabolic dish power plants have used linked dishes to create one or more heating loops to generate steam to drive a central steam turbine.

Solar Power Generation. DOI: http://dx.doi.org/10.1016/B978-0-12-804004-1.00006-3

The first parabolic dish solar devices were built by a French mathematician named Augustin Mouchot. In 1978, helped by his assistant Abel Pirfre, Mouchot presented his invention at the Universal Exhibition in Paris. The solar device consisted of a truncated cone-shaped reflector with an area of 20 m^2 that could produce steam pressurized to 3 bar to drive a pump capable of delivering 1500–2000 L/h of water. He is also credited with a solar refrigeration device. Mouchot's devices were never developed commercially, and the next time solar dishes appeared was a century later during the 1970s oil crisis. Some demonstration projects based on parabolic dishes were built during the 1980s, but no successful commercial plants. Since then a number of companies have continued to develop solar dish power units, and one plant of over 1 MW has been constructed. Even so, commercialization of the technology has yet to prove successful.

SOLAR DISH TECHNOLOGY

The solar dish is defined by the parabolic reflector that is used to capture and concentrate solar energy. The parabolic shape allows all the sunlight incident upon the reflecting surface to be focused to a single point where a heat collector or receiver is placed. The actual amount of heat energy collected will depend upon the size of the dish. Most modern dishes have been less than 10 m in diameter, although one project involves a 25 m diameter dish with an aperture or capture area of 500 m^2. A diagram of a solar dish power plant with multiple solar dishes is shown in Fig. 6.1.

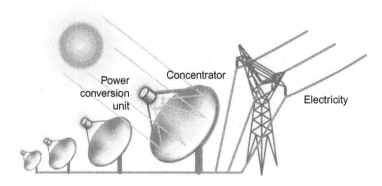

Figure 6.1 A solar dish power plant. Source: U.S. Department of Energy.

In most solar dish designs each dish has its own heat engine that is sized to suit the amount of solar energy the dish can collect. As a general guide, a 10 m solar dish can collect sufficient energy for a 25 kW heat engine.

The construction of the parabolic reflector is the most important aspect of the solar dish design. It is not economical to construct a single glass parabolic reflector of the size required for a typical solar dish, and so most dishes are broken down into a series of smaller elements that together form the larger reflector. Silvered glass is the most effective and durable material for mirrors, and offers one method of construction, with the individual mirrors mounted onto a parabolic space frame. For the largest parabolic dish yet built, the SG4 project in Australia, the reflector was constructed from identical laminated glass on metal mirrors that were mounted onto a tubular space frame structure. The reflector was constructed from 380 mirrors, each 1.165 m^2.

Other construction techniques have also been tried. Plastics are cheaper than glass, and the addition of a silver or aluminum layer to a transparent plastic support can provide a good reflective surface. However, plastics are likely to be less durable than glass. Another technique that has been tried is to stretch a flexible plastic membrane over a circular support, and then to use a partial vacuum behind the membrane to create the curved shape. This same technique has also been used to form a thin steel disk into a parabolic reflector.

The accuracy of the reflective surface is important for achieving high solar concentration levels. The dish must also be able to track the Sun across the sky. Two tracking systems have been used for solar dish projects. For larger dishes the azimuth-elevation system is preferred because it allows for better weight distribution. Under this system the disk can be rotated about two perpendicular axes, one horizontal and the other vertical. Rotation about the vertical axis changes the azimuth or compass bearing, while rotation about the horizontal axis alters the altitude or elevation. Smaller dishes often use polar tracking under which the dish moves about an axis running parallel to the Earth's axis of rotation and at a speed of 15°/h to match the speed of the Earth's rotation. There is also a second axis of rotation, perpendicular to this, to take into account the seasonal change in the height of the Sun in the sky.

POWER GENERATION IN SOLAR DISH SYSTEMS

The power generation system in a parabolic dish system normally needs to be a small, self-contained heat engine. In most cases one of two different types has been used, either a micro turbine or a Stirling engine. Steam generation, used to drive a steam turbine, is also possible, but is less common.

The key to the operation of each of these systems is efficient capture and transfer of heat energy to the thermodynamic working fluid that is used in the heat engine cycle. The heat capture and transfer is carried out using a receiver, which for a dish system may be either a tube-based receiver or a type of volumetric receiver. The heat can be transferred directly to the working fluid in the heat engine or it can be first captured by an intermediary heat-transfer fluid that then transfers the heat to the working fluid in a heat exchanger. Direct heating of the working fluid is the most efficient approach, and this is frequently used in dish systems.

Both direct and indirect heating can be applied to the Stirling engine. The latter is a closed cycle piston engine that contains either helium or hydrogen, pressurized up to 200 bar. Both gases are efficient at absorbing heat energy. The engine requires both a hot source and a cold source. The cold source is usually air from the atmosphere while the hot source is the heat from the solar collector. In the Stirling engine both heat and cold are applied to the outside of the engine. It is possible to heat a single engine directly by focusing the sunlight onto the heater plate of the engine. However, for more complex systems with multiple pistons it can be more efficient to use heat pipes to absorb the solar heat and transfer it to the places where it is required. Systems of this type have been built using liquid sodium as the heat-transfer fluid.

Stirling engines can be very efficient. With a solar input temperature of 1000°C, an efficiency of 40% is theoretically possible, although the highest efficiency that has been achieved so far is just over 31%. Even so, this is much more efficient than any other solar thermal technology. Typical Stirling engines for solar dish generation systems are between 10 kW and 30 kW in size.

Micro turbines are the other common small generating unit for solar dishes. A micro turbine is a small gas turbine, and its working fluid is air. Air is less good at absorbing heat than either hydrogen or helium, and to gain the best efficiency a type of volumetric receiver is required. This consists of a porous foam or honeycomb of a ceramic material. The receiver is exposed to sunlight through a quartz window, and the concentrated solar energy heats up the honeycomb. Air then becomes heated as it is pumped through the receiver.

The micro turbine has an open thermodynamic cycle. Air from the atmosphere is drawn into the turbine through its compressor stage, where it is compressed. This compressed air then passes through the volumetric receiver where it is heated to as much as 850°C. The hot, high-pressure air is then released through the turbine stage of the micro turbine where it generates sufficient power both to drive the compressor—mounted on the same shaft—and drive a small generator producing electricity. The still-hot air exiting the micro turbine exhaust is used to heat the air between the compressor and the volumetric receiver before being released back into the atmosphere.

The efficiency of a micro turbine is lower than that of a Stirling engine, but the units are cheaper. The micro turbine also has the advantage that supplementary heating can be added very simply by burning natural gas to boost output when the solar input falls off. As with Stirling engines, the typical size of a micro turbine solar dish plant is 10–30 kW.

In addition to the Stirling engine and the micro turbine, some solar dish systems have been designed to raise steam to drive a steam turbine. One of the earliest solar dish projects, built in southern California in 1983/1984, used a steam system. The project, called Solar Plant 1, used 700 dishes constructed from stretched-membrane reflecting elements. A system of steam loops, each one cycling through multiple dishes, was used to generate superheated steam at 490°C to drive a central, 4.9 MW steam turbine. The plant stopped operating in 1990.

More recently, solar dish development in Australia has also focused on steam generation. The SG4 large dish system is designed to generate steam directly. The 500 m^2 dish should be capable of driving a 100 kW steam turbine generator. In tests it has produced steam at 45 bar and 535°C using a tube boiler.

COMMERCIAL SOLAR DISH PROJECTS

There have been a number of proposals for large solar dish arrays, including some schemes involving thousands of dishes and generating capacities of hundreds of megawatts. So far, none has ever been built. However, two semi-commercial projects have been constructed, both in the United States. The first, called the Maricopa project, comprised 60 dishes, each of 25 kW, to provide an aggregate generating capacity of 1.5 MW. The project, which started generating power in 2010, was a demonstration scheme, and was intended as the precursor for a string of projects, including a 27 MW scheme with 1080 dishes and one with 34,000 dishes and a proposed generating capacity of 850 MW. Neither of these schemes, nor a third of 709 MW, ever materialized, and the Maricopa plant was decommissioned in 2011.

The second large-scale scheme was constructed at the Tooele Army Depot in the U.S. state of Utah. This project comprised 430 individual dishes, each capable of generating 3.5 kW of power for an aggregate capacity of 1.5 MW. Commissioning of the plant began in 2013, but later in the year the company that was building it filed for Chapter 11 bankruptcy. The fate of the project is unknown.

With no commercial projects under construction, solar dish technology has little chance of developing, even though it can offer high efficiency compared to other solar thermal technologies. The major difficulty has been competition from solar photovoltaic technology because the cost of the latter has fallen steeply. While other solar thermal technologies, and solar tower and solar trough plants, are suited to large-scale applications, the solar dish is by its nature a small-scale generating system. This has made it difficult for the technology to find a place in the renewable generation mix that is developing in the 21st century.

CHAPTER 7

Other Solar Thermal Technologies

In addition to the mainstream solar thermal technologies discussed in Chapters 4–6, there are two other approaches that exploit the heat in sunlight to generate electricity. These are the solar chimney and the solar pond. Neither has been deployed commercially. The solar chimney involves construction of a large greenhouse that acts as a solar collector, heating the air within the structure. This hot air is funneled toward the center of the structure to a large chimney through which the hot air ascends, creating a powerful updraft. This updraft of air is then used to drive wind turbines to produce electricity. The solar pond, meanwhile, is a low technology approach to generating electricity. It involves a large pond that absorbs heat from the Sun, leading to the creation of a temperature difference between its upper and lower layers that can be used to produce electricity from a small heat engine.

THE SOLAR CHIMNEY

The solar chimney concept involves the construction of a large, normally circular, open-sided greenhouse. The glass cover of the greenhouse retains the air that is heated by sunlight during daylight hours. The roof of the greenhouse slopes slightly upwards toward the center of the structure, encouraging the hot air to flow in that direction.

At the center, the greenhouse structure is punctured by a tall chimney. The hot air flows up this chimney, creating a powerful updraft. At the same time, more air is drawn into the greenhouse from the open sides. The current of air created by this structure can be used to drive wind turbines. These turbines can either be placed within the chimney itself, or more economically, they can be placed at ground level around the perimeter of the chimney. Energy storage can be added to the scheme by installing bags containing water on the floor of the greenhouse. These heat up during the day and release the heat at night, maintaining the air flow to drive the turbines.

Solar Power Generation. DOI: http://dx.doi.org/10.1016/B978-0-12-804004-1.00007-5

This concept has been tested at a small scale. However, commercial plants based on the idea would need to be massive, and this has so far prevented a commercial project from being commissioned. A schematic of a solar chimney power plant is shown in Fig. 7.1.

The single solar chimney pilot scheme so far constructed was developed in Spain by a German company called Schlaich Bergermann Solar. The prototype was built between 1981 and 1982 at Manzanares and was funded by the German government. This pilot scheme had a tower 190 m high and 10 m in diameter. It was surrounded by a glass collector with a radius of 122 m. A turbine was installed at the base of the tower and was able to produce an output of 50 kW. The project operated until 1989. The company is still promoting the technology with proposals for plants of 200 MW in generating capacity and chimneys of up to 1 km in height.

Various other companies and organizations have taken up the solar chimney concept. These include an Australian company that proposed building a 200 MW scheme in the state of Victoria in 2004. Two African countries, Namibia and Botswana, have also taken an interest in the concept, and a small test facility was operated briefly in

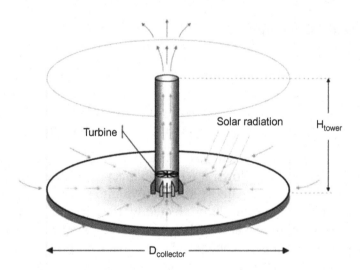

Figure 7.1 A solar chimney power plant. Source: Schlaich Bergermann Solar Gmbh.[1]

[1]http://www.solar-updraft-tower.com/en#technical_concept/principle.

Botswana in 2005. Meanwhile, the Namibian government approved a scheme to build a 400 MW tower, 1.5 km high, with 37 km^2 of greenhouse that would have been used to grow crops as well as generate electric power. In spite of these ambitious schemes, no large-scale solar chimney has been built.

THE SOLAR POND

The solar pond is a low-technology solution to solar heat collection for power generation. The concept upon which it is based has been tested at several locations, but it has never proved commercially viable.

The solar pond consists of a large, watertight excavation or container. The pond, usually around one to two meters deep and several thousand square meters in surface area, is filled with a brine made from salt (sodium chloride) and water. The surface of the brine is usually covered with an impermeable transparent material to prevent evaporation. The pond, once established, is exposed to sunlight, which heats up the brine.

Under normal circumstances, with fresh water, a pond heated in this way would establish a vertical thermal gradient so that the hottest water was at the top of the pond and the coolest at the bottom. However, the brine causes an inversion so that a hot layer of concentrated brine collects at the bottom of the pond while a layer of cooler, much less concentrated—and hence less dense—brine forms at the top. In between these two a gradient zone is established in which both temperature and salinity change. This helps keep the more dense, hot brine at the bottom.

The vertical temperature gradient within the pond can be used to drive a heat engine. To achieve this, hot brine is withdrawn from the bottom of the pond and used to heat and evaporate a thermodynamic working fluid. The gaseous fluid is then used to drive a turbine, creating a pressure gradient across the turbine by condensing the working fluid using cooler brine from the top of the pond.

The temperature gradient from top to bottom of the pond is likely to be no more than 50°C at best, with the hot brine reaching up to 80°C while the surface temperature is around 30°C. The best way to exploit this small temperature difference is by using an organic Rankine cycle turbine system. This is a closed cycle system based around a small turbine that is like a steam turbine. The working fluid in the cycle is a low boiling point organic fluid that both boils and condenses within the temperature range of 30°C to 80°C.

Apart from its simplicity, the solar pond has the added advantage of a built-in heat storage system. Depending on the size of the pond relative to the turbine used to extract energy, the complete system should be able to generate power around the clock.

SOLAR POND SCHEMES

Large solar pond projects have been limited. One of the most significant was developed close to the Dead Sea in Israel during the early 1980s. The pond was constructed by excavating a cavity and lining it with rubber sheets to create a surface of 7000 m². Once established, the temperature at the bottom of the pond stabilized at between 70°C and 80°C while the temperature at the surface was between 20°C and 30°C. This pilot scheme was able to generate 150 kW of power using an organic Rankine cycle turbine built by a company called Ormat, which developed the scheme. Subsequent to this a 5 MW project was built at Beit Ha'Arava and operated until 1988. The pond for this scheme had a surface area of 210,000 m² and overall efficiency was around 1%. The eventual aim of these projects was to harness the Dead Sea for power generation, but the scheme was apparently dropped by the beginning of the 1990s.

Another, smaller solar pond was constructed at the University of Texas, El Paso. The pond was a converted fire protection pond and had an area of 3000 m². The project operated intermittently between 1986 and the mid-1990s. A project was also completed in Bhuj in the state of Gujarat, India, in 1993, but this only supplied hot water.

CHAPTER 8

Solar Photovoltaic Technologies

Of all the forms of renewable power generation, the solar cell is perhaps the most elegant. The cell is a solid-state device so it has no moving parts; it simply exploits the physical properties of the elements from which it is made to capture sunlight and convert it into electricity. Although it is based on extremely sophisticated technology, a solar cell is easy to deploy. Size, both large and small, is no constraint. At their smallest, solar cells can be used to provide power to tiny consumer devices such as calculators or watches. At their largest, massive arrays are used to create grid-connected power plants that can generate tens or even hundreds of megawatts of electric power. The efficiency is the same in each case.

Solar cells (or more precisely, solar photovoltaic devices, because they rely on the photovoltaic effect) are manufactured using the same type of technology used to make transistors and microchips. While the technology is costly, economies of scale have allowed the cost of the solar cell to fall to such a level that it is becoming competitive with other forms of power generation. The key transition point will be reached when solar cells reach economic parity with mainstream generating technologies. When that will happen is a subject of hot debate, but researchers will certainly be approaching that point by the end of the second decade of the 21st century. Indeed, in some situations and regions, the economic parity barrier has already been breached.

The simplicity and ease of deployment of this technology means that over the long-term this is probably the most promising means of replacing large-scale fossil fuel-based power generation with a renewable resource. That may take several decades to achieve, but with a global total of 40 GW of new capacity installed in 2014 (see Table 1.1), and an aggregate global capacity of close to 180 GW at the end of that year, the process is well underway.

Solar Power Generation. DOI: http://dx.doi.org/10.1016/B978-0-12-804004-1.00008-7

ORIGINS

The history of the solar cell begins with the discovery of the photovoltaic effect by French scientist Alexandre Edmond Becquerel in 1839. Becquerel was experimenting with an early battery comprising two metal electrodes in an electrolyte solution. During this work he observed that the amount of electricity generated increased when the electrodes were exposed to light. Becquerel's observation was followed in 1873 by the discovery that the electrical conductivity of selenium increased when it was exposed to light, results that were published by the British engineer Willoughby Smith. Finally, in 1883, American inventor Charles Fritts coated selenium with a thin layer of gold to create the first photovoltaic cell. It had a light-to-electrical energy conversion efficiency of about 1%.

Although these advances were used in photocells and other simple light-sensitive devices, it was the discovery of the p-n semiconductor by Russell Ohl in 1939 that lead to the development of the first modern solar cell, based on a silicon p-n junction. This revolutionary silicon solar cell was demonstrated at Bell Labs in the United States by Daryl Chapin, Calvin Fuller, and Gerald Pearson in 1954. The first devices achieved about 6% efficiency, much higher than had been possible with simple selenium photocells. The new silicon solar cells were used in the U.S. Vanguard satellite in 1958.

Early photovoltaic cells were expensive. Consequently, while they were used in space programs where cost was no object but reliability was vital, they found little application elsewhere. Prices remained high during the 1960s and 1970s, but the evolution of methods to grow large single crystals of silicon for use in integrated circuits gradually brought the price of the raw material down, and the price of silicon solar cells started to fall.

During the 1970s and 1980s solar cells were still considered expensive, but they were more readily available and they found a use in niche markets such as navigational buoys and remote telecommunications stations. As production volumes increased and costs continued to fall, the number of applications increased. However, it was not until the 1990s that large-scale application of solar cells for power generation began to advance, particularly in domestic and commercial rooftop applications.

While single crystal silicon was, and remains, the most important base for solar cells, other technologies have also been developed. Polycrystalline silicon became popular during the 1990s because it was much cheaper to produce and appeared capable of challenging single crystal silicon in performance. Even cheaper thin film amorphous silicon solar cells were also developed. During the first decade of the 21st century, thin film solar cells based on a number of other materials, cadmium telluride in particular, attracted attention and appeared to offer an extremely cheap and practical alternative to silicon. However, the cost of silicon cells continued to fall, and their high efficiency, coupled with low price, has allowed their continued domination of the market.

There is also a niche market for high-efficiency solar cells, some based on more complex silicon solar cell structures, others exploiting materials like gallium arsenide and indium telluride, which have an inherently higher efficiency. However, they are generally only used in specialized applications such as concentrating solar cells.

SOLAR PHOTOVOLTAIC BASICS

Modern solar cells are made from a variety of different semiconductors, all of which have the ability to absorb light from the visible spectrum. The operation of these solar cells depends upon a fundamental property of these semiconductors called the bandgap, a part of the quantum-level structure of the material relating to the distribution of electrons within the solid.

The bandgap is a function of the specific semiconductor's electron energy levels, which results in an energy gap between the top layer of full electron energy levels and the first set of empty energy levels. In a conductor, this empty band of energy levels is so close in energy to the full band below it that electrons can easily jump from one to the other as a result of thermal activation. The electrons in the upper level can then move easily across the material in the mostly empty energy band of energy levels, conferring electronic (or at the macroscopic scale, electrical) conductivity on the material.

In a semiconductor the energy gap is too large for electrons to jump from the lower to the upper level as a result of thermal activation at

Table 8.1 The Bandgap of Some Common Solar Cell Semiconductors	
Semiconductor	Bandgap (eV)
Silicon	1.11
Cadmium telluride	1.44
Gallium arsenide	1.43
Copper indium gallium diselenide	0.9–1.7

normal temperatures. However, electrons in the lower level can become promoted from the lower to the upper level, across the bandgap, by absorbing photons of electromagnetic radiation. For this to be possible, the photon must contain at least as much energy as the size of the energy gap between the two sets of energy levels in the semiconductor.

The range of electromagnetic radiation that the cell can absorb is determined by the size of its bandgap. Semiconductors that are useful for solar cells have bandgaps that make them capable of absorbing photons within the visible region of the solar spectrum. All those with an energy greater than the bandgap can be absorbed.[1] However, any photon with an energy lower than the bandgap (such as infrared radiation) will not be absorbed. Table 8.1 shows the bandgaps of some semiconductors commonly used for solar cells. The bandgap of silicon is 1.11 eV.

Each photon of light energy absorbed by the semiconductor is captured by an electron within the material. In absorbing the energy, the electron acquires an electrical potential relative to the electrons around it because it has a higher energy. The special structure of a photovoltaic cell created by the p-n junction allows this potential to be exploited to provide an electric current. The current is produced at a specific fixed voltage called the cell voltage. The cell voltage is, again, a property of the semiconducting material. For silicon it is around 0.6 V.

The energy contained in light increases as the frequency increases from infrared through red to blue and ultraviolet light. However, a solar cell must throw away some of these frequencies since it can only absorb light above its cell threshold, defined by the semiconductor

---

[1]The energy of a photon in electron-volts is given by $E = hf$, where h is Planck's constant and f is the frequency.

bandgap. Light that is of an energy below this threshold simply passes through the material.

It might seem sensible, therefore, to use a semiconductor with a low threshold or bandgap. However, this would lead to a cell with a low output voltage because these factors are also directly related to the threshold for absorption. There is another drawback to using a semi-conductor with a small bandgap. When a photon of light with energy much greater than the threshold energy is absorbed, it loses all the energy above the threshold value. The surplus energy is essentially thrown away (it emerges as heat) and cannot be used for electricity production.

These two factors mean that the lower the bandgap and absorption threshold, the greater the number of photons absorbed but the more energy thrown away as heat; the higher the bandgap, the more energy simply passes through the material without being absorbed. The optimum conditions are therefore a compromise between the two competing effects.

The optimum bandgap for a solar cell semiconductor is 1.43 eV. As Table 8.1 shows, this is exactly matched by gallium arsenide. However, this material is much more expensive than silicon, and the presence of arsenic has environmental implications. The bandgap of silicon, at 1.11 eV, is less than optimum, but it has proved to be the most effective solar cell material to date, and has the largest market share. Silicon has been used in three different forms: crystalline, semi-crystalline, and amorphous. Crystalline silicon is the most efficient, but also the most expensive to produce, while amorphous silicon is both the least efficient and the cheapest. Alternatives to silicon include cadmium telluride, which is cheaper to produce and is always in amorphous form. Its bandgap is close to optimum. Meanwhile, copper indium gallium diselenide has a bandgap that varies with composition, allowing a degree of tailoring. Once again, however, the material is much more expensive than silicon.

CHAPTER 9

Types of Solar Cells

Solar cells are manufactured using technologies similar to those used to produce microchips and transistors. Today most of these are made using slices of perfect silicon crystals. The slices are then chemically etched and doped to create the complex structures required for computers and other electronic devices. A solar cell, though simpler in structure than a microchip, can be manufactured in a similar way.

Silicon solar cells made from single crystal silicon are the most efficient available, with reliable commercial cell efficiencies of up to 20%. The record laboratory efficiency for a single solar cell under normal solar irradiation conditions is 25%. Even though this is the most expensive form of silicon, it also remains one of the most popular due to its high efficiency and durability. Polycrystalline silicon is cheaper to manufacture, but the penalty is lower efficiency, with the best measured around 18%. This is also very popular. Cheapest of all to produce is amorphous silicon, which can be made using vapor deposition techniques rather than by expensive crystal growing. However, its best efficiency is only 8%. Amorphous silicon also suffers from degradation when first exposed to light, a problem not seen with crystalline material. This can reduce its initial efficiency by up to 20%.

All silicon solar cells require extremely pure silicon. The manufacture of pure silicon is both expensive and energy intensive. The traditional method of production required 90 kWh of electricity for each kilogram of silicon. Newer methods have been able to reduce this to 15 kWh per kilogram. This still means that depending upon its efficiency, a silicon solar cell can take up to 2 years to generate the energy needed to make it. This compares with around 5 months for a solar thermal power plant. Manufacturers of crystalline silicon are concentrating on ways of reducing the cost of crystalline material by cutting it more efficiently, by reducing the amount in each solar cell, or by finding new ways of growing it. This is helping push costs down, and crystalline silicon

Solar Power Generation. DOI: http://dx.doi.org/10.1016/B978-0-12-804004-1.00009-9

remains competitive in spite of efforts by thin film manufacturers using much cheaper materials. Meanwhile, manufacturers are also exploring ways of increasing the average lifetime of a solar cell. Modern cells are usually rated for a 25-year life, but it appears possible with suitable encapsulation techniques to increase this to as much as 100 years.

Another crystalline material used for solar cells is gallium arsenide. This has an almost perfect bandgap for a solar cell, and in the laboratory efficiencies of 28% have been recorded. However, practical cells only reach 20%, and the material is both expensive and composed of hazardous materials. It is rarely used, except for special applications.

The main alternative to crystalline silicon for solar cells is some form of thin film. From a manufacturing point of view these are attractive because they can be produced using cheap techniques such as vapor deposition, sputtering, or even printing. Amorphous silicon is one alternative, but it is not as cheap to produce as cadmium telluride (CdTe), and the latter has a much higher efficiency, with the best recorded at 22% (though the efficiency of the best commercial cells is only 15%). This material also has an almost optimum bandgap for a solar cell (1.44 eV), and its potential efficiency could approach 30%. CdTe is also attractive because it is possible to produce solar cells on a variety of substrates, including building components and flexible plastic sheets.

The maximum efficiency possible with a single-layer solar cell of any semiconductor is 33.7%. It is possible to build more efficient solar cells by layering cells one on top of the other. Such multilayer or multi-junction cells place the semiconductor with the largest bandgap at the top. This top layer absorbs high-energy radiation, but lower-energy radiation passes through it to reach the layers below, where further semiconductor layers of lower bandgap are placed. In principle it is possible to create a cell with up to around 50% energy efficiency with multilayer cells, but these cells are much more expensive to manufacture. The best recorded efficiency achieved to date is 43.7%.

SILICON SOLAR CELLS

While there are a range of solar cells made from different materials, the most common as a proportion of solar cell production are cells made from crystalline silicon. These have accounted for between 80% and 90% of global production since around 1992. Two types dominate—single crystal silicon and polycrystalline silicon—with each accounting for around 40%. The single crystal material is made from high purity silicon using the Czochralski process, in which a single crystal is pulled slowly from a crucible of molten silicon. Polycrystalline silicon is composed of a multitude of tiny single crystals and is much simpler to make, though it too requires high purity silicon at the outset.

Trace amounts of other materials can be added to the molten silicon to change its properties. The two common types are n-type doping in which a small amount of an element such as phosphorous or arsenic (with more outer electrons than silicon) is added to the mixture, creating some additional electrons that can move more freely than those in the undoped silicon. The alternative p-type doping involves adding an element with fewer outer electrons, such as boron or gallium. This leads to "holes" being created in the lower band of electron energy levels; these holes can behave like positive electrons, again moving freely. Both types of doping lead to increased conductivity in the silicon.

The key structure for the solar cell is a p-n junction. This is created by taking a single crystal of silicon that has been doped to produce one of these types and then using a technique to change the type of doping in a part of the crystal so that a junction between the two types of doping is created across adjacent layers of the material; the same effect can be achieved by carefully growing a layer of alternative doping on top of the original. Where the two types meet there is a concentration gradient within the bands of the semiconductor because one region naturally holds more electrons while the other has more holes. The natural tendency in such a situation is for electrons to move into the region where there is a surplus of holes and vice versa. The result is that a charge gradient is set up across the junction, which eventually counterbalances the concentration gradient and prevents more particles from moving, as shown in Fig. 9.1.

It is this in-built charge gradient that makes the solar cell function. When the solar cell material absorbs a photon of light, this promotes an electron from the lower energy band to the upper energy band, across the bandgap in the semiconductor. The electron is now free to move, and since it has a negative charge it travels toward the positively

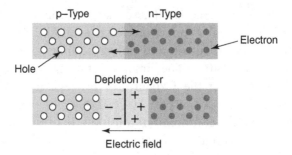

Figure 9.1 A solar cell p-n junction showing the charge gradient. Source: Wikipedia commons.[1]

charged side of the junction while the positively charged hole left behind travels in the opposite direction. The moving charges will create a current if the cell is connected to an external circuit.

The conventional structure for a solar cell is planar. The cell is made of a thin layer of semiconductor, the top surface of which is doped with a suitable impurity to create the p-n junction within the surface layer of the material. The junction, with its built-in voltage gradient, will then capture electrons once they are generated by light absorption and sweep them away into an external circuit. In order to collect the current, electrical contacts must be formed onto the semiconductor. The rear of the cell can be covered with a planar collector since light does not have to pass through this surface, but on the front surface, where light is absorbed, the collector area must be minimized or it will interfere with light absorption. The front collector is usually made from narrow fingers of metal that allow the maximum amount of sunlight to strike the semiconductor surface. The semiconductor and its contacts will often be placed on a substrate to give it additional strength, and the top layer will be covered with a transparent protective coating. A schematic of a crystalline solar cell is shown in Fig. 9.2.

More advanced structures are possible. The front contacts can be buried, edgewise, into the surface of the cell to minimize the shadow effect of the contact, which reduces efficiency. More modern cells have also been developed that move both contacts to the back of the cell by allowing the front doping to be carried to the back of the semiconductor. This structure is more complex, but creates a more efficient cell by removing any shadowing from the front contacts.

[1]https://commons.wikimedia.org/wiki/File:PnJunction-PV-E.PNG.

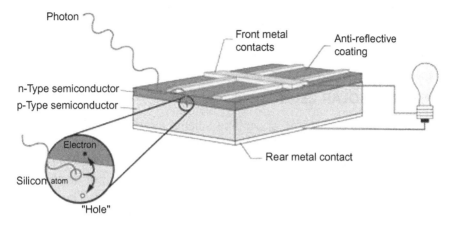

The overall efficiency of a solar cell depends on effective light absorption. This can be improved by making the front surface of the cell non-reflecting and by making the back reflective so that any light reaching the back layer is reflected back into the semiconductor. All these improvements further increase the overall complexity of the cell. Such developments mean that the number of stages in the fabrication of a solar cell can increase to perhaps 15 from a more typical 6−8 in the simplest devices. However, decreasing production costs and increasing efficiency are making such developments worthwhile.

The first solar cell ever produced was made using n-type silicon as the substrate, but over recent years most silicon solar cells have been made from p-type silicon onto which the n-type layer is added to create the p-n junction. More recently, interest has again risen in n-type silicon as the main substrate upon which the solar cell is built. There appear to be advantages because n-type silicon is often of higher quality and performs better in mass production environments. These n-type solar cells have shown high efficiency both in laboratory tests and in solar cell modules.

While laboratory tests have been able to achieve around 25% efficiency in a silicon solar cell, the theoretical maximum efficiency for this type of cell is 29.4%. Techniques such as better light management to improve absorption may be able to push the efficiency higher.

[2]http://www.viridiansolar.co.uk/Solar_Energy_Guide_5_2.htm.

However, to go much higher than today's best will require multilayer or tandem cells in which layers of different semiconductors are used to absorb more of the solar spectrum.

THIN FILM SOLAR CELLS

The main competition for crystalline silicon solar cells comes from a variety of thin film solar cells. Manufacturing solar cells from thin film materials is much easier than making them from solid semiconductor substrates because they can either be produced by vapor deposition techniques, by sputtering, or in some cases by printing. The film is deposited onto a substrate that is often glass, but may also be plastic. The nature of the manufacturing technique means that much larger area cells can be made than are possible with crystalline silicon. This too has economic advantages because fewer cells have to be interconnected, making module production simpler. In addition, it is possible to produce thin film solar cells on flexible materials, including fabrics that might be used for clothing.

There are several thin film semiconductors that have been developed for solar cells. Amorphous silicon is a form of silicon with no crystal structure that is produced using thin film techniques. Its efficiency is low compared to the crystalline material, but it is much cheaper to manufacture and it continues to command a share of the thin film market, particularly for small solar cells for electronic devices.

The main alternative is CdTe. This too is cheaper than crystalline silicon, and solar cells made from it can be deposited on a variety of substrates. It is also significantly more efficient than amorphous silicon. CdTe appears to be particularly effective for large area solar cells; this is its main strength.

Another material that has been developed for thin film solar cells is copper indium gallium diselenide ($CuInGaSe_2$, sometimes known as CIGS). By varying the amounts of copper, indium, and gallium, the bandgap of this material can be changed, and this can be used to tailor the thin film for a specific application. However, production costs appear to be higher than for CdTe. A variant of CIGS is copper indium diselenide. This is also being developed for thin film applications.

The manufacture of a typical thin film CdTe solar cell starts with a perfectly clean glass substrate.[3] Onto this is first sprayed a Transparent Conducting Oxide (often referred to as the TCO) layer. The conducting layer can be made from one of a variety of materials, including zinc oxide, indium oxide, indium tin oxide (ITO), and strontium oxide. The conducting layer allows light passing through the glass to reach the semiconductor below, while also providing a top contact to extract power from the cell. On top of the conducting layer, the next layer to be deposited is one of n-type cadmium sulfide that forms the first part of the p-n junction. This is usually referred to as the window layer because light has to pass through it to reach the semiconductor below. Then a layer of p-type CdTe is added. Finally, a conducting back contact is deposited onto the CdTe layer. This is usually a metallic layer such as nickel—aluminum. Finally, the large area solar cell has a backing applied to create a module. A cross section of a CdTe solar cell is shown in Fig. 9.3.

Although there will be variations in the techniques for depositing the various layers, and in the composition of the layers themselves, the process for manufacturing thin film cells from any suitable semiconductor will involve similar stages. Meanwhile, research continues to optimize the various layers, particularly the window layer, in order to achieve the highest efficiency possible.

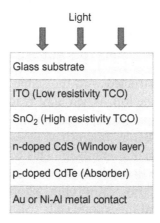

Figure 9.3 Cross section of a cadmium telluride (CdTe) thin film solar cell. Source: Wikipedia.[4]

---

[3]Details are based on the manufacturing process used by First Solar.
[4]https://commons.wikimedia.org/wiki/File:Cadmium_telluride_thin_film_solar_cell.png.

One drawback of very large-scale deployment of CdTe or $CuInGaSe_2$ is the availability of the elemental constituents. Tellurium is a by-product of copper refining. While it is readily available today, the U.S. Department of Energy has predicted a shortfall by 2025. Indium is also available in limited quantities via the refining of a variety of metals such as zinc, copper, iron, and lead, and is in demand for liquid crystal displays and a variety of coatings. (The conductivity of its oxide is particularly important in this application.) The main producer is China. Availability is not considered an issue today, but could become so in the future. In contrast, silicon is one of the most abundant elements on the planet.

MULTILAYER SOLAR CELLS

One way of improving the efficiency of electricity production based on solar cells is to use multiple solar cells, with each one capable of absorbing a part of the solar spectrum. In this way more of the energy from the Sun can be utilized. In order for this to be effective, the different solar cells must operate in tandem; these devices are sometimes called tandem cells. The components of a tandem cell are normally built one on top of the next.

As already noted, the absorption of light by a solar cell is controlled by the bandgap of the semiconductor from which it is made. Each semiconductor absorbs photons of light with energy greater than its bandgap, but remains transparent to those photons with energy lower than its bandwidth. These simply pass through the cell.

A tandem cell is constructed with the semiconductor solar cell with the highest bandgap at the top. When light reaches the cell, the highest energy photons are captured by this solar cell, but lower energy photons pass through. Below this cell is a second one made from a semiconductor with a smaller bandgap. This cell absorbs photons that have passed through the first cell. It is possible to have a third layer to capture even more energy, absorbing photons that have passed through both the first and the second cells.

Typical cells of this type might use gallium indium phosphide to capture the highest energy photons at the blue end of the visible spectrum. Gallium phosphide could then be used to capture light photons from the yellow and green parts of the spectrum while a silicon cell would be used to capture low energy visible light photons and infrared photons.

In theory, a multilayer solar cell with an infinite number of layers could achieve close to 87% energy conversion efficiency. In practice it is unlikely that more than two or three layers will be used. Such cells have shown a laboratory efficiency of 43%. Commercial two-layer cells have shown 30% efficiency under normal sunlight conditions and up to 40% when exposed to concentrated sunlight.

The multilayer technique has also been applied to create amorphous silicon solar cells with improved efficiency. One process involves building a cell that is composed of a mixture of layers of amorphous silicon and microcrystalline silicon. This helps with light absorption and therefore improves efficiency. A more cost-effective alternative is to build a multi-junction solar cell comprising a series of layers of amorphous silicon. Very thin amorphous silicon solar cells appear to suffer less degradation over time, however, the thin layer does not absorb all of the light and a significant part passes through. By using multiple layers, more light can be captured while retaining the resistance to degradation. It is also possible to modify the properties of some of the amorphous silicon layers with doping to adjust the bandgap to capture more infrared light.

ENHANCED LIGHT ABSORPTION TECHNIQUES

Effective absorption of light is key to the efficiency of a solar cell. If all the light is not absorbed then energy is being wasted. Although this can be an issue with all types of solar cells, it is most often discussed in relation to silicon cells.

One way of ensuring that most of the light is absorbed in a silicon solar cell is to use a thick layer of silicon. However, this has an impact on the cost of the cell because more high purity silicon means higher material costs. The trend is therefore toward much thinner cells. If the layer of silicon is too thin, then light will pass all the way through it without being absorbed.

Designers have adopted several strategies to counter this. One is to make the light bounce backwards and forwards within the layer of silicon until the photons have been absorbed. To achieve this, the back surface of the silicon must be made highly reflective so that any light reaching it is reflected back into the cell. Meanwhile, the top surface of the cell is coated with a layer that causes light to be reflected internally again rather than escaping.

Preventing external reflection before light enters the cell is important too. The top outer surface of a silicon solar cell is normally relatively reflective so that around 20–30% of the light that strikes it from outside is reflected rather than allowed to pass into it. Another advanced technique involves etching the surface in such a way that a structure of tiny needles around 10 microns in height and 1 micron or less in diameter is produced. Known as black silicon, this surface treatment can reduce the reflectivity so that only around 5% of the incident light is lost through reflection.

Another advanced technique that can potentially increase the efficiency of all silicon solar cells further is called upconversion. As with all semiconductors, silicon can only absorb photons that have an energy greater than its bandgap. All those longer wavelength photons with a lower energy cannot be absorbed. In upconversion, some of these lower energy photons are absorbed by a special dye. The dyes that are used for upconversion have the ability to absorb two or more low energy photons and then release one photon at a higher energy. So long as this new photon has an energy larger than the semiconductor bandgap, it can now be absorbed. This technique is still at the laboratory stage, but it promises the ability to utilize a wider range of photons. It could raise the maximum theoretical efficiency of a single layer solar cell from around 30% to 50%. The technique could prove of particular value with amorphous silicon solar cells.

CONCENTRATING SOLAR CELLS

The concentrating solar cell offers a slightly different approach to the capture of solar energy using solar cell technology. While traditional solar cells are large area energy capture devices that utilize sunlight at the intensity at which it arrives from the Sun, as well as capturing some diffuse radiation, the concentrating solar cell adopts the approach of concentrating solar thermal power plants.

Concentrating solar systems rely primarily on direct irradiation from the Sun. They cannot concentrate diffuse radiation. This means that most must employ some form of tracking system if they are to be effective. In addition, they are only cost-effective if they are deployed in regions with high direct insolation levels.

A concentrating solar cell uses a cheap concentrating or focusing device to capture solar energy over a wide area and then focuses it onto a small area solar cell. For this to be effective, it requires first that the large area concentrating system be cheaper than an equivalent large area planar solar cell module. Secondly, it requires a solar cell that can absorb the highly concentrated solar energy effectively. Since this solar cell is tiny compared to normal large area planar cells, it is cost-effective to use more complex and expensive devices. Some of these, such as tandem cells, have a higher efficiency than traditional planar devices.

Concentrating solar cells have a technical advantage over planar cells because the efficiency of a solar cell depends on the intensity of the light that is incident upon it. Even a simple planar cell will show higher efficiency if the solar energy is concentrated. More complex multilayer solar cells are capable of even higher efficiency, as noted above. The best efficiency so far demonstrated with a multilayer solar cell is around 45%, and practical modules are expected to be able to deliver DC current at around 36% efficiency. This is twice the efficiency of the best planar solar cells operating under normal light conditions.

How a concentrating solar cell is deployed, and the type of cell used, depends on the concentration level being achieved with the optical system. Low concentration systems use simple, cheap, plastic lenses to generate a solar intensity of between 2 and 100 Suns. The concentration of the solar energy focuses the heat energy from the Sun as well as the light energy, which leads to a significant heat buildup in the solar cell onto which the energy is directed. For low concentration devices, normal silicon solar cells can be used without the need for active cooling, although the design of the system may involve passive cooling of the cells. In addition, such low concentration systems can be deployed without the need to track the Sun across the sky, yet they are still capable of up to 35% efficiency.

For higher concentration levels, both cooling and active tracking are needed. This increases the complexity of the system and increases the overall cost, but can result in greater efficiency. Medium concentration systems that concentrate solar energy by 100–300 times can also use silicon solar cells, but they may exploit alternatives such as gallium arsenide cells or multi-junction solar cells. With plastic lenses and efficiency of up to 40%, this type of system could be cost-effective in a range of applications, including rooftops.

High concentration solar photovoltaic devices usually use similar technology to that used in solar thermal plants, particularly the dish-type system. With a parabolic dish concentrator, a concentration ratio of 1000 times can be achieved. The solar cells and electronic systems in these devices need to be cooled actively, and need to be capable of withstanding relatively high temperatures. The solar cells used are likely to be high-efficiency multilayer cells that, while much more expensive than a simple silicon cell, can achieve significantly higher efficiency. Production cells can now reach 44% efficiency.

An advanced form of concentrator that is still at the development stage is a device called a luminescent concentrator. This comprises a transparent plastic plate that incorporates a luminescent dye compound[5] or, alternatively, the luminescent material may be contained in a thin film that coats the planar surfaces of the plate. When this plate is exposed to sunlight, the luminescent dye absorbs sunlight and then re-emits its own luminescent photons. The collector is designed in such a way that the re-emitted fluorescent light is channeled within the transparent plate in much the same way as laser light is trapped inside a fiber optic cable. In the same way as light emerges from the ends of a fiber optic cable, so the fluorescent light emerges from the edges of the plate. Solar cells placed around the edges capture this light to produce electricity. The fluorescent plate collector has the advantage over more traditional concentrators in that it can capture diffuse as well as direct radiation.

THIRD-GENERATION SOLAR CELLS

When solar cells are divided into generations, silicon and other single crystal-based solar cells are considered to be the first generation. Continuing on from these, thin film solar cells may be thought of as

---

[5]As well as luminescent dyes, other options include semiconductor fluorescent quantum dots that confine electrons in a three-dimensional space, and compounds containing rare Earth metals.

the second generation. There is now a third-generation made up of solar cells that use either organic semiconductors or dye-sensitized semiconducting materials to capture light and convert it into electricity. These materials are nowhere near as efficient as the earlier generations of materials, and probably never will be. Their advantage is that they can be manufactured extremely cheaply and produced on a diverse range of different substrates; this opens up the possibility of their use in niche products from architectural elements to clothing and consumer electronic devices.

There are a number of organic polymer semiconductors that have been developed in recent years. These can be used to produce light-emitting diodes and printed transistors. Development of small electronic devices using these materials has been successful, but for large-scale energy capture there are difficulties to be overcome. The main obstacle is that an organic polymer semiconductor is a poor conductor of electricity. This means that when a photon is absorbed by the material, generating a free electron and a hole, the two do not get swept away from one another as they would in a material such as silicon, but stay in the same place. This leads to rapid recombination of electron and hole, and the energy is lost as heat.

To overcome this, the organic semiconductor must be combined with a second material that preferentially captures the electron or hole (or both) and allows them to separate before they can recombine. In approaching this issue, scientists have been studying photosynthesis, where a similar process is achieved in organic materials. Current efficiency of organic semiconductors is around 5% at best. The near-term target is to achieve 10% by using advanced designs.

Dye-sensitized solar cells also exploit principles found in plants and photosynthesis. In this case it is a dye that absorbs the photon of light. The cell is created by absorbing the dye onto tiny particles of a semiconductor: titanium oxide. A layer of these particles is then coated onto one of the solar cell electrodes. Above the coated electrode is a conducting electrolyte that is capped with a second transparent electrode and a glass cover. This is shown schematically in Fig. 9.4. When a photon is absorbed by the dye, the electron created is captured at the dye—titanium dioxide interface, while the hole travels through the conducting electrolyte, thus ensuring that they separate before they can

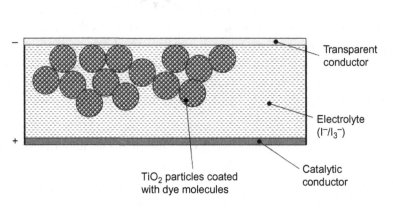

Figure 9.4 A schematic of a dye-sensitized solar cell. Source: Wikipedia.[6]

recombine. Efficiency is generally higher than for organic semiconductors, and can reach 13% in actual devices.

Both organic and dye-sensitized solar cells can be printed, so manufacturing is cheap. They are also capable of operating in low-light situations. However, they are not as stable as more traditional devices, and can be damaged by ultraviolet light.

[6]https://commons.wikimedia.org/wiki/File:Dye_sensitized_cell_phys.jpg.

CHAPTER 10

Modules, Inverters, and Solar Photovoltaic Systems

The solar cell is only a part of a complete solar photovoltaic (pV) package. Individual silicon solar cells are relatively small in terms of both physical size and the amount of power they can produce. In order for them to be integrated into power supply systems, large numbers of individual solar cells are aggregated by connecting them both in series to increase the output voltage, and then connecting series strings in parallel so that power can be increased. The aggregated cells form a single module that must then be encapsulated so that it is protected from the elements, fitted with electrical connections so that it can be connected into the power system, and attached to a mounting system (Fig. 10.1).

Individual thin film cells are often much larger than silicon cells so that each cell has a relatively larger output, and in the case of cadmium telluride solar cells, one cell can form a single module. However, it too must be encapsulated with a protective enclosure and provided with connections. The same applies to concentrating cells, but these are often more complex because they comprise both a solar cell and a light-concentrating system.

The output of a solar module is a direct current (DC) at a DC voltage determined by the size of the module. For some applications, such as in isolated supplies where there is no grid connection, it is possible to use this DC supply directly to supply power to electronic equipment, for lighting or heating, or to charge a battery system that will provide power when there is no sunlight. However, in most applications the solar pV system will be integrated with a grid supply, and this means that the DC supply must be converted to grid frequency alternating current (AC). This is achieved with a device called an inverter, which for most systems today is based around solid-state power electronic components. In the case where the solar module is part of a grid-connected solar pV power plant, there may be hundreds or thousands

Solar Power Generation. DOI: http://dx.doi.org/10.1016/B978-0-12-804004-1.00010-5

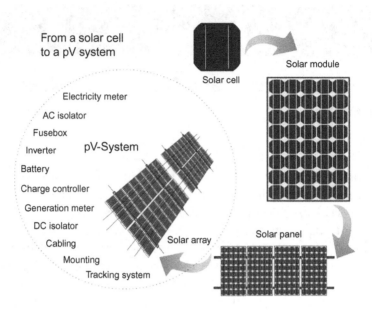

Figure 10.1 Solar cells to solar panels. Source: Wikipedia.[1]

of modules, and the power from all of them must be collected and converted into a form that can be delivered into the grid.

The precise nature of this balance of plant components will depend on the type of deployment. A small domestic rooftop installation might involve a handful of modules that are mounted directly onto the roof and equipped with a relatively simple inverter and grid interface. At the other end of the scale a large, grid-connected solar pV power plant will have all its solar modules mounted onto supports with tracking systems so that the solar cells can follow the Sun across the sky. There will be a substation providing the grid interface and power quality control, and the plant may provide grid support services too.

CELL AND MODULE OUTPUTS

The output of a solar cell or solar module varies with the load that is applied to it. When it is open circuit, the output voltage will be at its highest but it will be supplying no current. When a load is applied to the output, the output voltage will fall as the current increases. Initially the voltage will fall gradually, and linearly, as the current increases from zero. Then, at a certain inflection point the rate at

[1]https://commons.wikimedia.org/wiki/File:From_a_solar_cell_to_a_PV_system.svg.

which the voltage falls will increase. This inflection point is called the operating point of the cell, and it is the point at which the cell provides its maximum power (see Fig. 10.2). The power at this point is called the "peak power output," and this figure is used as the rating of the solar cell or module. Peak power output is affected by the temperature of the cell, and at higher temperatures it falls off slightly.

A typical silicon solar cell will have a surface area of $15 \, cm^2$. A commercial cell of this size will have an output of 6 A at 0.55 V, or a peak power output of around 3 W. Module sizes vary, but a commercial rooftop module might contain 60 individual, interconnected solar cells providing an output of 250 W at peak output, at a voltage of around 30 V and a current of 8.3 A. Typical outputs for commercial panels vary between 175 W and 315 W, although most are around 200−220 W.

Table 10.1 compares the characteristics of solar modules made from different types of solar cells. The module efficiencies follow the material

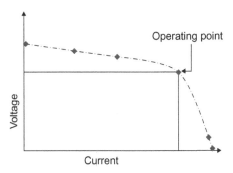

Figure 10.2 Solar cell output characteristics. Source: REUK website.[2]

Table 10.1 Characteristics of Solar Photovoltaic Modules		
Cell Material	**Typical Module Efficiency Range (%)**	**Typical Output per Square Meter (W)**
Monocrystalline silicon	15−18	110−140
Polycrystalline silicon	13−16	110−130
Amorphous silicon	6−8	50−80
Cadmium telluride	9−11	80−90
Copper indium diselenide	10−12	90−110
Source: Energypedia.[3]		

[2]http://www.reuk.co.uk/Measuring-the-Power-of-A-Solar-Panel.htm.
[3]https://energypedia.info/wiki/Solar_Cells_and_Modules.

efficiencies already discussed in earlier chapters, so that single crystal (monocrystalline) silicon modules are the most efficient, with a typical efficiency range of 15−18%. These cells generally produce 110−140 W for each square meter. Polycrystalline modules are slightly less efficient at around 13−15%, and produce 110−130 W for each square meter. Meanwhile, amorphous silicon modules have a much lower efficiency, 6−8%, and generate only 50 W to 80 W/m^2. Cadmium telluride modules are more efficient than this, with an efficiency range of 9−11% and an output of 80 W−99 W/m^2. Meanwhile, copper indium diselenide has an efficiency of 10−12% in module form and produces 90 W−110 W/m^2.

INVERTERS

In some situations it is possible to take the DC output from a solar panel and use it directly to provide power. For example, the panel might provide a dedicated power supply to a specific piece of electronic equipment. Virtually all modern electronic equipment requires a DC supply, and if supplied from the grid, the AC supply must be converted to DC with a consequent energy loss, so a direct DC supply can be more efficient. Further, a solar module could provide power directly to charge a battery. Solar modules destined for this type of stand-alone operation can be sized to match the needs of particular battery or electronic equipment.

This type of direct DC connection from solar module to power consumer is feasible, but in most situations it is actually more efficient to use an interface between the solar module and the equipment or battery being supplied. This interface, which is designed to optimize power capture from the solar module under varying light conditions, is called an inverter.

An inverter takes the power from the solar pV device and conditions it for use. In a typical stand-alone application an inverter might control the output of a solar module being used to charge a battery. In this case it will take a varying DC output from the array of solar cells and convert it into a stable DC supply. To do this, the DC power is converted into a high-frequency AC supply and then back to a stable DC voltage suitable for the battery.

More often a solar power system will be required to provide an AC output at the grid frequency so that the solar system can be integrated into a grid-supplied system. In this case the inverter takes the DC output from the solar panel or from a set of solar panels and converts it into an AC output. In doing so the inverter has to cope with a continuously varying output from the solar devices as the solar conditions vary, and yet still produce a stable output at grid frequency and at grid voltage.

There are a range of different types of converters available to suit different applications. These include stand-alone inverter systems where the solar output is not connected to a grid-supplied system. Meanwhile some hybrid inverters send DC power to a battery bank to keep an energy storage system charged while the remainder of the power from the solar system is used to supply power to the grid or to a domestic or commercial establishment that is grid-connected. Finally, the most common type of inverter today only provides an output that is grid-connected and grid-synchronized.

The inverter itself is an electronic module that uses solid-state power devices to condition the power output from the solar panels. For smaller domestic and commercial solar pV installations there are two types of inverters in common use. One type, called a micro inverter, handles the power from an individual solar panel. The second type, called a string inverter, takes its input from a number of panels. Each has its advantages and disadvantages. The string inverter is the most well-established type of inverter for solar pV, and it is the cheapest and potentially the easiest to install. However, each string inverter must handle the power from multiple panels so they must have a high power capacity, and this can affect the cost and reliability of the components from which they are made. In addition, a string converter treats all the panels connected to it in the same way. The inverters use an adaptive technology to try and obtain the maximum output from the solar panels, but if one or more of the panels is shaded or soiled, or if a panel fails, this can affect the performance of all the panels. String inverters also have limited flexibility when it comes to expanding a solar array as they are normally sized for the initial number of panels. This makes it difficult to add more panels at a later stage.

Micro inverters are designed to manage the power from a single solar panel. There will therefore be a lot more of them in a system that involves more than one or two panels, and they will usually be more expensive than a single string inverter. However, since each panel is managed independently, the effect of the shading of one panel, or a panel failure, will be confined to that panel. Micro inverters also handle, individually, much lower levels of power than string inverters, and can be made from potentially smaller and more reliable electronic components. Advocates claim that they are capable of improving overall efficiency of a solar power system by between 5% and 20%. Micro inverters are potentially more flexible than string inverters because new panels can easily be added to expand a system. Some solar panel manufacturers are beginning to supply panels with integrated micro inverters, making installation even simpler.

Utility-scale solar pV power plants generally use large string inverters because the cost of micro inverters would be too high; their advantages are much less clear. Utility plants normally have tracking solar modules and the units are cleaned regularly to maintain performance. In addition, they are sited so that modules are never in the shade, except as a result of the Sun setting. Therefore, being able to isolate the performance of each panel is not so important. Expanding a utility plant would involve the addition of multiple modules too, so there is no advantage to be gained from being able to expand one module at a time. While inverters for rooftop applications may be able to handle several kilowatts, for utility plants the inverters have capacities of up to 500 kW each. In a typical plant, individual inverters deliver their power to external transformers that feed into a medium-voltage power collection system connected to a central substation where a substation transformer steps the power up to grid voltage before delivery into the system.

MAXIMUM POWER POINT TRACKING

In order to get the best from solar cells and solar panels, most modern inverters use a technique called maximum power point tracking (MPPT). The technique is employed to optimize power capture from solar cells under varying conditions. The output of a solar cell varies depending on the intensity of the light striking it, the angle of incidence at which the light strikes the cell, and the temperature of the cell. For each set of conditions, that solar cell has a specific maximum power point, a specific voltage and current for which its power output is at its maximum. MPPT invokes systems in the inverter that are able to sense how the actual output of the cell compares to its optimum as the conditions vary, and then adjusts its own operating conditions so that the cell can produce its maximum power.

Some of these MPPT inverters use simple approaches such as a fixed output voltage to control the solar cells. However, others use a much more sophisticated approach such as "perturb and observe," where the output conditions (current and voltage) of the cell are altered slightly at regular intervals, by the inverter, and the power measured; if the new power is higher, then the conditions are moved further in that direction until the power starts to drop off, when the optimum point has been found.

Gains from MPPT will vary. However, it may be possible to increase performance by 10–20% during the summer and 30–40% during the winter by using the technique.

GRID INTERFACING

While some solar pV systems operate independently of any grid system, the majority of pV systems in use are grid-connected. Large- and medium-sized utility power plants based on solar pV generation feed all their power directly into the grid, and they have to be able to meet grid codes of performance when they do so. However, many solar pV systems are much smaller rooftop systems that supply power either to individual dwellings and to commercial operations, or operate as distributed generation systems supplying power to a group of small users. In each of these cases the system will usually have a dual role, to supply power to its local consumers and to supply power to the grid. These, too, have to meet grid standards.

Grid connection is controlled by a stringent set of codes and conditions to ensure that a system connected to the grid does not degrade the grid supply by affecting grid voltage, grid frequency, or by adding unwelcome noise or harmonics to the grid. Depending on its size, a solar installation may also need to be able to provide support for the grid, either by being able to increase or reduce its output on demand from the grid controller, or by helping stabilize the grid voltage and frequency. Further, any pV system connected to the grid will have to respond in a controlled way to the grid power supply failing, a situation known as islanding.

The inverters used to convert the DC output from a solar panel into grid frequency AC are capable of controlling the frequency and voltage they produce for the grid as the input voltage from the solar system varies. The control systems are often fast acting and their output stable. This type of output can help strengthen the grid and reinforce it against fluctuations caused elsewhere in the grid. However, for this to be effective, there needs to be some form of communication between the power supply unit and the central control room that manages the grid.

Communication systems of this type are relatively expensive and are not yet widespread. If such communication systems are established, it becomes possible to modulate the output of the solar unit to suit grid needs. If this is coupled with accurate weather forecasting, which helps predict how solar output will vary during the day—potentially on an hourly basis—then even rooftop solar installations can be used for grid support services as well as supplying power. This will become more important as the proportion of solar pV on the grid increases in coming years.

It is also important that power systems connected to the grid behave in a specific and predictable way when the grid supply fails. If the main supply trips as a result of a problem somewhere across the system, a pV system will normally continue to generate power. If this pV system is connected to the grid, then it will behave as an island of live power when the rest of the grid has lost power. This can cause problems on the grid side when technicians are trying to rectify the fault, and on the power plant side if it is feeding power into a dead grid. To avoid this, grid codes usually require that a solar pV system that is grid-connected be switched off or disconnected from the grid in the event of a grid-side failure.

SOLAR PHOTOVOLTAIC INSTALLATIONS

Solar cells can be used in a wide variety of situations, and this flexibility distinguishes them from any other form of power generation. If one excludes the smallest applications, in consumer electronic devices, for example, the application of solar pV is usually broken down into three or four categories. The main groups are distributed solar pV—including the many rooftop installations now in use—utility power plants, and off-grid applications. The distributed category can be further sub-divided into domestic rooftop pV, commercial and industrial rooftop installations, and some commercial and industrial ground-mounted solar pV facilities.

The sector that drove solar pV during the 1990s and the first decade of the 21st century was the domestic and small commercial rooftop use of solar panels. In several countries this was supported with govern-ment grants, and these programs led to a massive expansion in the use of solar pV. The most important of these early programs were in Japan—where a program ran from 1994 until 2005—and in Germany and the United States. Subsequently, incentive schemes of various types were launched in many countries, and many of these continue today. However, the level of support is beginning to tail off as the cost of solar cells falls to a level where subsidy is no longer needed.

The early incentive programs were vital in driving up the volume of solar cell production and bringing the price down to a level at which they now approach grid parity. In 2010 residential solar pV installa-tions accounted for 63% of global solar pV capacity according to the International Energy Agency (IEA).

Commercial and industrial solar pV represent a much smaller part of the total installed capacity. These installations are generally nothing more than larger versions of the typical domestic rooftop installation, although some of the largest are approaching utility-scale. Trends include companies covering the roofs of their distribution centers with solar panels. At the end of 2010 commercial installations accounted for 11% of global capacity.

Off-grid use embraces a wide range of applications. There are, for example, many different types of stand-alone installations, from air pollution monitoring stations to traffic monitoring and remote relay stations, that rely exclusively on solar cells for their power. There are also large numbers of people, particularly in the developing world, who have no access to grid power. Solar pV installations are increas-ingly being used to supply power to off-grid communities, and as the cost of solar cells falls, the economics become easier. There are also a wide range of individual dwellings in the developed world that are still remote from the grid, or choose to remain off-grid and use solar power, perhaps with wind power, to supply electricity.

While all these sectors continue to be of importance, the fastest growing sector in the second decade of the 21st century has probably been utility-scale power plants based on solar cells. The first mega-watt-scale plant of this type was the Lugo plant in California with a generating capacity of 1 MW. It started generating in 1982. The Lugo project was an outlier as was a 6 MW plant that was built in the United States in 1985. After these two, large-scale pV plant construction was abandoned until early in the first decade of the 21st century because the economics were not sustainable. Since around 2003 the capacities of both individual plants and the aggregate global utility plant have risen dramatically. The world's first 10 MW plant was erected at Pocking in Germany in 2006. A plant with a nominal capacity of 97 MW was built in Canada in 2010, and in 2011 a 200 MW plant was built in China. More recently, a 550 MW plant was built in California in 2013, and in 2015 this was overtaken by a 579 MW plant, also in California. The total capacity of utility-scale solar pV power plants at the end of 2014 was 36 GW according to Wiki-Solar, or roughly 20% of the total global solar pV installed capacity.

As solar pV capacity continues to grow, the main market segments will remain the same, although proportions may change. The IEA has predicted that in the period from 2014 to 2040, utility-scale and rooftop systems will each take roughly half the market.[4]

---

[4]Technology Roadmap: Solar Photovoltaic Energy 2014 edition, International Energy Agency, 2014.

CHAPTER 11

Solar Integration and the Environmental Impact of Solar Power

Solar power is one of the major renewable sources of electric power available on Earth, alongside hydropower and wind power. Hydropower is well-established and has the largest installed capacity of any renewable source. Rapid development of wind power began in the early 1990s, and its global capacity is now large too. Solar power developed alongside wind power beginning in the 1980s, but for most of that period it has been considered too expensive to offer an economical source of electricity. That changed following a dramatic fall in the cost of solar cells; solar capacity is rising rapidly now too. Over the long term, solar power, particularly in the form of solar cells, offers the most reliable and accessible form of renewable power generation. It is available globally, and during the rest of this century it is expected to develop into the most important source of energy.

The value of these renewable resources is their ability to generate electricity without emitting any carbon dioxide, and in most cases, any other pollutants. Global warming is now a major threat to global security and the global economy, so reducing these emissions has become a high priority. Renewable electricity generation is one of the most important tools available for achieving this.

In this context, solar power is not entirely clean. Production of silicon used to manufacture solar cells is an energy intensive process, and much of the energy used today to prepare pure silicon comes from power stations burning fossil fuel. Other solar photovoltaic (pV) cell semiconductor materials also have significant energy costs. In addition, solar thermal power plants require high-technology components such as mirrors and steelwork, and these require energy to manufacture. However, both types of power plants are responsible for fewer emissions over their lifetimes than any type of fossil fuel plant. The materials used in some solar cells are also hazardous, and that must be taken into

Solar Power Generation. DOI: http://dx.doi.org/10.1016/B978-0-12-804004-1.00011-7

account when considering the environmental impact of solar power. Overall, however, the environmental benefits far outweigh the risks.

Solar plants have other environmental effects. Land usage is high compared to fossil fuel plants, and solar thermal plants often need water, a problem in the arid regions to which they are best suited. There is also an operational issue with solar power, a consequence of the fact that solar energy is only available intermittently. Solar power can only be harvested when the Sun is shining. When the Sun goes behind the horizon, solar power plants stop producing power. This means that a solar power plant cannot provide a reliable supply of electric power on its own. This intermittency is a problem that rises in importance as the amount of solar power connected to the grid increases. Techniques to help mitigate intermittency problems are being developed alongside other advances in solar power generation.

THE ENVIRONMENTAL IMPACT OF SOLAR POWER

From an environmental impact perspective, solar thermal and solar pV power plants need to be considered separately. Solar thermal power plants are similar to traditional fossil fuel-fired power plants with regards to their use of steam turbines and associated equipment to generate power. However, solar plants use a large solar collector field instead of a coal boiler to collect heat and raise steam. This collector field is of significant size for a large power plant, using between 2 ha/MW and 5 ha/MW. Further, if land is used for solar energy capture it is difficult to use it for any other purpose, unlike wind power, where land can be used for crops or animal grazing. On the other hand, it is possible to build solar plants on poor quality land; the best sites are likely to be in desert regions where insolation is high. This makes it less likely that solar plants will be competing with other land users.

Of greater concern is water usage. A solar thermal plant requires some form of cooling in order to condense steam as part of the power generation heat engine cycle. This is normally provided by water cooling. A typical solar thermal plant with a recycling water cooling system requires around $2-3\,m^3$ of water for each megawatt-hour of power generated. Once-through cooling systems can use more water, although overall water loss through evaporation is generally lower. The alternative is dry cooling, which can cut water use by 90%, but at a cost to efficiency. However, dry cooling can be less effective at high ambient temperatures, conditions that are likely to be common where solar thermal plants are sited. Because the best sites for solar thermal power generation are arid regions where there is low rainfall and a generally dry climate, water use can present significant difficulties. It may be possible to bring water from less arid regions, but that would push up the cost, possibly making a plant uneconomical.

A large solar thermal power plant will affect the habitat upon which it is constructed. However, the effect may be relatively benign unless the site is also the habitat for an endangered species. The power plant collectors will offer additional shade that was not previously available, and this could prove beneficial. A potentially more serious issue relates to the effect of large solar tower plants on birds. These power plants function by achieving high levels of solar concentration, much higher than in other types of solar thermal plants. In essence, they create high temperature beams of sunlight that can injure or kill birds that fly through them. The issue is emotive and the severity of the problem is still under investigation, particularly in the United States, where the largest plants of this type have been built.

The construction of a large solar power plant, thermal or pV, involves local disruption of traffic movements and noise. There is also the possibility of potential spillages from the heat collection circuits and the steam generation components of a solar thermal plant. However, these factors should not represent a serious risk at a well-managed plant.

Like solar thermal facilities, utility-scale solar pV power plants have high land usage requirements. Typically a plant will use between 1 ha/MW and 4 ha/MW, slightly lower than for a solar thermal plant. As with a solar thermal plant, it is unlikely that the land can be used for any other purpose. This should not be a problem if plants are sited in areas where the soil is poor and has little agricultural value. Aside from the disruption during the construction phase, a large solar pV plant should have little local impact.

Land use is much less of an issue with smaller solar pV installations as many of them are placed on rooftops. These installations can often be unsightly, but in most instances they are not visually intrusive. The development of architecturally designed rooftop panels can certainly improve their appearance, as can the integration of solar panels at the design stage in building construction. Some modern developments include solar panels as part of the structure; these usually stand out less, and may even prove visually appealing. There have been proposals to incorporate solar cells into windows too. Based on luminescent collectors, these could potentially make solar generation invisible. However, this technology still has a long way to go before it can be deployed commercially.

Other than land usage, solar pV plants have little impact on their environment while operating. Utility plants have mechanical systems to allow the modules to track the Sun, and these could cause minor oil spillages. Rooftop systems, however, have no moving parts. Problems could potentially arise when a plant reaches the end of its life. Decommissioning these plants involves recycling large quantities of semiconductor material. Regulatory approval of many modern pV plants requires the manufacturer of the solar pV cells to take them back for recycling at the end of the life of the plant. This is particularly important for some thin film cells that include environmentally hazardous materials.

LIFE CYCLE COST

While solar power plants produce electricity without the emission of carbon dioxide, they are not emission free. The components of both solar thermal and solar pV power plants are made using energy intensive processes. These include the production of steel, glass, and specialty metal components for the heat absorbers in a solar thermal plant, as well as all the components of the steam turbine generator plant. Significant quantities of concrete—an energy intensive product—are likely to be used in both types of solar plant. In addition, there is one outstanding life cycle cost for silicon pV plants, the energy cost involved in the production of the silicon.

The actual cost in emission terms of producing silicon depends on the mix of technologies that are generating the electricity that is used to manufacture the silicon. For the situation in Europe, estimates suggest that the lifetime carbon emissions of a solar pV power plant are between 20 gCO_2/kWh and 80 gCO_2/kWh. This is around 10 times less than the emissions from a fossil fuel power plant.[1] These emissions are equivalent to an energy payback time for a commercial plant in Southern Europe of 0.7–2.5 years. The life cycle cost for cadmium telluride can be of the same order of magnitude. This compares to lifetime carbon emissions from a coal-fired power plant of around 900 gCO_2/kWh, and emissions from a natural gas-fired power plant of 400 gCO_2/kWh. Solar thermal power plants show better lifetime performance than solar pV plants, with average lifetime emissions of around 30 gCO_2/kWh.

These figures depend in part on the lifetime of the actual power plant. The assumed lifetime of a solar pV plant is 25 years. If this was extended to 50 years or even 100 years, a lifetime that may be possible in the future, then the lifetime emissions would fall. In addition, as the world switches from fossil fuel to renewable, the lifetime emissions will fall because renewable sources will contribute more of the power that is used during the manufacturing processes.

SOLAR INTEGRATION

Solar power is by its nature both intermittent and unpredictable. The source is intermittent because the Sun only shines during daylight hours. This has to be taken into account when building grid-connected solar stations.

The change from night to day and back is highly predictable, as are the seasonal changes as the Earth moves around the Sun. This predictability makes it relatively easy for grid controllers to manage the broad cycle of solar power generation. However, on top of the diurnal variation in solar intensity there is a random variability resulting from the weather. Clouds can reduce solar intensity significantly and rapidly, and this can lead to big changes in output from solar power plants.

The variability of solar output is not an issue when there is only a small amount of solar power being injected into a grid system compared to the overall grid-connected generating capacity. All grids are designed to cope with variations in both demand and supply. However, when the level of solar penetration as a proportion of total production on the grid rises beyond a certain point, the solar input can

[1]Technology Roadmap: Solar Photovoltaic Energy 2014 edition, International Energy Agency, 2014.

become a problem because it behaves differently to other sources. This is complicated by the fact that solar power must normally be dispatched when it is available, so grid operation must be designed around maximum use of solar—and other—intermittent renewable sources.

The variations in solar power output from solar plants connected to a grid system can be substantial. Across the nations of the European Union in 2011, the average amount of solar power available for dispatch during high summer reached close to 10,000 MW. In the depth of winter it was only 2200 MW. Short-term variations can be massive too. This is particularly noticeable at sunrise and nightfall when the output from solar plants ramps up and down very rapidly. Meanwhile, the arrival of cloud cover can reduce solar output by as much as 50% over the area affected. In 2011 in Germany, the maximum hourly change in pV output was 10,300 MW; in Italy it was 7200 MW. Grid operators have to be able to manage these changes by using other plants to support the solar output.

At a grid-wide level, resilience in the face of large amounts of solar power input depends upon factors such as the amount of energy storage or fast-acting grid capacity that can step in or back out as solar input rises and falls. Grids that have significant amounts of hydropower can be particularly resilient because hydroplants can start up and shut down very quickly. At the grid line level the problem often manifests itself as one of voltage stability. This is frequently caused by too much power being injected into the distribution system to which many rooftop solar systems are connected. The excess power then feeds back from the distribution system to the transmission grid through the distribution/transmission system transformer substations. In most conventional grids power is only supposed to flow from the transmission system to the distribution system, not in the other direction.

The problem of solar input may not be grid-wide, however. It can cause local problems too, on a single grid line, for example. Again, the most pragmatic way of determining the effect is through grid voltage. When this starts to fluctuate as a consequence of solar input, the solar penetration is too high. At what penetration level this occurs depends on the situation. It might start at 15% solar penetration if the grid line is long and has little voltage and frequency support; on a stronger line it might appear until penetration reaches 95%.

There are mitigating factors that help accommodate solar power at all levels of the grid. One of the most important factors is the coincidence of peak solar output with the hottest period of the day. This occurs when there is the greatest demand for air conditioning, so solar output will, at least partially, match a particularly hot summer day's demand and alleviate the need for grid controllers to bring expensive peaking power plants online.

Another important means of managing solar power is by using accurate weather forecasting. Modern forecasting techniques can provide a good level of reliability, and forecasts are used routinely in most grid control centers today. Good day-ahead and hour-ahead forecasting, particularly if used to predict the weather over each solar plant connected to the system, can make it easy for operators to anticipate changes in solar power output and plan for other sources to either back out or come online as needed.

New technology can also help manage solar input into the grid from solar pV systems. A device that is at the forefront here is the smart inverter. As noted in Chapter 10, inverters form the interface between the solar pV system and the grid. A modern inverter can be programmed to modulate the output of the solar system it controls. In the past this has been carried out simply by switching the pV output on or off. However, new inverters can carry out partial "curtailment" by reducing the output rather than shutting it off completely. If the installation is able to communicate with the grid controller, this can be controlled centrally, allowing a fine level of management.

The final grid tool for managing intermittent sources such as solar power is energy storage. Energy storage can be implemented at the transmission grid level, at the distribution grid level, or at the level of individual pV installations. There are many storage technologies that can be used, but the most important for solar pV are batteries, often installed at a local level. These are fast acting and can either absorb power or supply it, depending upon the conditions. At the grid level, pumped storage hydropower is also effective. Energy storage is considered expensive, but it is likely to form a much more significant component of national grids as renewable generation grows.

CHAPTER 12

The Cost of Solar Power

The cost of electricity from a power plant of any type depends on a range of factors. First there is the cost of building the power station and buying all the components needed for its construction. In addition, most large power projects today are financed using loans, so there will also be a cost associated with paying back the loan, with interest. Then there is the cost of operating and maintaining the plant over its lifetime. Finally, the over-all cost equation should include the cost of decommissioning the power station once it is removed from service. It would be possible to add up all these cost elements to provide a total cost of building and running the power station over its lifetime, including the cost of decommissioning, and then dividing this total by the total number of units of electricity that the power station produced over its lifetime. The result would be the real life-time cost of electricity from the plant. Unfortunately, such a calculation could only be completed once the power station was no longer in service. From a practical point of view, this would not be of much use. The point in time at which the cost-of-electricity calculation of this type is most needed is before the power station is built. This is when a decision is made to build a particular type of power plant, typically based on the technol-ogy that will offer the lowest cost electricity over its lifetime.

LEVELIZED COST OF ENERGY MODEL

In order to get around this problem, economists have devised a model that provides an estimate of the lifetime cost of electricity before the station is built. Of course, since the plant does not yet exist, the model requires a large number of assumptions and estimates. In order to make this model as useful as possible, all future costs are also con-verted to the equivalent cost today by using a parameter known as the discount rate. The discount rate is almost the same as the interest rate and relates to the way in which the value of one unit of currency falls (typically, but it could rise) in the future. This allows, for example, the maintenance cost of a solar thermal power plant 20 years into the

Solar Power Generation. DOI: http://dx.doi.org/10.1016/B978-0-12-804004-1.00012-9

future to be converted into an equivalent cost today. The discount rate can also be applied to the cost of electricity from the solar power plant in 20 years' time.

The economic model is called the levelized cost of electricity (LCOE) model. It contains a lot of assumptions and flaws, but it is the most commonly used method available for estimating the cost of electricity from a new power plant.

When calculating the economics of new power plants, the levelized cost is one factor to consider. Another is the overall capital cost of building the generating facility. This has a significant effect on the cost of electricity, but it is also important because it shows the financial investment that must be made before the power plant generates any electricity. The comparative size of the investment needed to build different types of power stations may determine the actual type of plant built, even before the cost of electricity is taken into account. Capital cost is usually expressed in terms of the cost per kilowatt of generating capacity in order to allow comparisons between technologies.

When comparing different types of power stations, there are other factors that need to be considered too. The type of fuel, if any, that it uses is one. A coal-fired power station costs much more to build than a gas-fired power station, but the fuel it burns is relatively cheap. Renewable power plants can also be relatively expensive to build. However, they normally have no fuel costs because the energy they exploit is from a river, from the wind, or from the Sun, and there is no economic cost for taking that energy. Once the renewable power plant has been paid for, the electricity it produces will have a low cost. All these factors may need to be balanced when making a decision to build a new power station.

THE CAPITAL COST OF SOLAR POWER PLANTS

The capital cost of a solar power plant depends on a number of factors, the most important of which is the cost of the components needed to build the plant. For a solar power plant, these components include the collection field with its mirrors and tracking systems, heat exchangers, the steam generator and steam turbine, and ancillary equipment. In this type of plant the collector field can account for up to 50% of the total cost, depending upon the type of plant. For a solar photovoltaic (pV) plant the solar modules will probably account for close to half the cost, while the balance of the plant will account for the rest.

The cost of the components needed to construct the plant and the labor required to build it can be rolled up into a figure called the overnight capital cost, which excludes the cost of any loans needed to finance the building of the plant. The overnight capital costs of both solar thermal and solar pV plants in the United States are shown in Table 12.1. The figures in the table are from the U.S. Energy Information Administration's (EIA's) Annual Energy Outlook for each year in the table; the figure refers to the cost calculated for the year previous to the publication year.

For solar thermal plants the estimated cost in the 2001 report was $3681/kW. At that time there were no major U.S. solar thermal plants in operation. The estimate fell dramatically in 2003 to $2204/kW, before rising to $2675/kW by 2007. There was a steep jump in 2009, coinciding with a sharp rise in commodity prices. After some variations during the financial crisis, the capital cost from the 2015 report was estimated to be $3787/kW.

The estimated capital cost of solar pV power plants has shown similar variations. From $2394/kW in the 2001 report, prices rose to $5750/kW in 2009, but fell back sharply to $3123/kW in the 2015 report. These U.S. EIA solar pV costs are significantly higher than figures from some other sources. Lazard,[1] for example, put the cost in

Table 12.1 Capital Cost of Solar Power Plants in the United States, 2001–15		
Report Year	Capital Cost of Solar Thermal Power Plant ($/kW)	Capital Cost of Solar pV Plant ($/kW)
2001	3681	2394
2003	2204	3389
2005	2515	3868
2007	2675	4114
2009	4693	5750
2011	4333	4474
2013	4653	3624
2015	3787	3123
Source: U.S. Energy Information Administration Annual Energy Outlook 2001–15.		

[1]Lazard's Levelized Cost of Energy Analysis—Version 9.0, Lazard 2015.

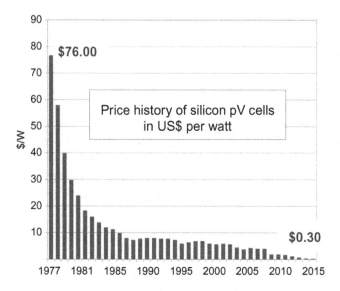

Figure 12.1 The price of silicon solar cells, 1977–2015. Source: Wikepedia.[2]

2015 of utility solar pV at $1600–1750/kW. The cost of rooftop systems was higher.

Predicting the cost of solar pV power plants is difficult because a large part of the price depends on the cost of solar cells or solar modules, and the cost of these has been dropping dramatically, and continues to drop. This is illustrated in Fig. 12.1, which shows how the price of silicon solar cells has varied since 1977. Based on the data in the figure, the cost in 1977 was $76.00/W. In 2015 it was estimated to be $0.30/W. As manufacturing capacity continues to rise and technology improves, the price is likely to drop even further.

THE LEVELIZED COST OF SOLAR POWER

For solar power plants the main contribution to the levelized cost of power comes from the capital cost. In addition, there are ongoing maintenance costs for both types, as well as the cost of financing any loans.

Table 12.2 shows the levelized cost of solar power in the United States for a number of different solar configurations, based on estimates from Lazard. The levelized cost of residential rooftop installations is

[2]https://commons.wikimedia.org/wiki/File:Price_history_of_silicon_PV_cells_since_1977.svg.

Table 12.2 Levelized Cost of Power from Solar Power Plants in the United States	
	Levelized Cost of Electricity ($/MWh)
Residential rooftop solar pV	184–300
Commercial rooftop solar pV	109–193
Utility-scale solar pV (crystalline solar cells)	58–70
Utility-scale solar pV (Thin film solar cells)	50–60
Solar thermal power plant with energy storage	119–181
Source: Lazard's Levelized Cost of Energy Analysis—Version 9.0, Lazard 2015.	

$184–300/MWh, and for commercial and industrial rooftop installations it is between $193/MWh and $109/MWh. While these costs are relatively high compared to some of the others in the table, it is important to remember that these installations are competing with the retail cost of power to the residential or commercial consumer. This will be much higher than the average wholesale cost of power to the grid.

For utility-scale solar power plants the levelized cost is much lower. For a plant that uses crystalline silicon solar cells the levelized cost is $58–70/MWh, while for a plant with thin film solar cells the levelized cost is $50–60/MWh. These cost ranges put solar pV in competition with wind power and gas turbine combined cycle plants as among the cheapest sources of power in the United States. Solar cells are traded globally, so costs will be broadly similar across the globe.

Solar thermal power is much more expensive. From Table 12.2 the typical range of levelized cost is $119–181/MWh for a solar thermal plant with energy storage. The cost for a plant without storage (not shown in the table) was estimated to be $251/MWh. Based on these figures solar thermal power is similar in cost to that produced by offshore wind farms.

One key question is when solar power will reach parity with other forms of power generation. On the basis of the figures in Table 12.2, solar pV power appears to be close to parity in the middle of the second decade of the 21st century. The cost of power from competitive forms of power generation will vary from region to region, so it is not possible to make any sweeping claims based on these figures. However, it seems clear that parity is not far away.

Printed in the United States
By Bookmasters